ALLERGIC 2 MATHS

MAKE PEACE WITH MATHS

Volume 1

Decimals, relative numbers, squares and square roots, fractions and powers of numbers.

Pascal IMBERT

Copyright © 2015 Pascal Imbert

Table of contents

INTRODUCTION   **7**

INTEGERS AND DECIMALS NUMBERS   **9**

AMERICAN VOLUME UNITS   10
A LITTLE ASTRONOMY   13
THE OLYMPIC GAMES   15
CALCULATE DECIMALS   **18**
TO THE MARKET   18
HOW TO CONVERT IN DIFFERENT UNITS?   **21**
TO THE MARKET (CONTINUED)   21
HOW TO SUBTRACT DECIMALS?   **24**
TO THE MARKET (CONTINUED)   24
HOW TO MULTIPLY DECIMALS?   **26**
TO THE MARKET (CONTINUED)   26
HOW TO DIVIDE DECIMALS?   **28**
THE MOVE   34
THE STAMP COLLECTOR   35
GOOD ACCOUNTS MAKE GOOD FRIENDS   38
THE DIFFICULT CHOICE OF TILES   41

FRACTIONS   **45**

WHAT IS A FRACTION?   **46**
THE KING OF COCKTAILS   46
CALCULATE WITH FRACTIONS   **50**

HOUSE FOR SALE   50

WE TAKE CREDIT?   51

THIS IS A SURVEY   53

COMPARE FRACTIONS   **55**

THE LEGACY OF AN ECCENTRIC UNCLE   55

SIMPLIFY FRACTIONS   **59**

LET'S PLAY SCRABBLE®!   59

HOW TO MULTIPLY FRACTIONS?   **63**

THE GROCER   63

HOW TO ADD FRACTIONS?   **65**

THE BEST PASTRY   65

A GENEROUS FRIEND   67

RELATIVE NUMBERS   69

CLASSIFY AND CALCULATE THE RELATIVE NUMBERS   **70**

HISTORY AND TIMELINE   75

SCUBA DIVING   77

SQUARES AND SQUARE ROOTS   80

WHAT IS THE SQUARE OF A NUMBER?   **80**

ROAD SAFETY   80

WHAT IS THE SQUARE ROOT OF A NUMBER?   **83**

ROAD SAFETY (CONTINUED)   83

POWER OF A NUMBER   86

WHAT IS A POWER?   **87**

POWERS OF 10   **88**

ASTRONOMY   88

HOW TO MULTIPLY BY A POWER OF 10?   **90**

ASTRONOMY (CONTINUED)   90

WHAT IS SCIENTIFIC NOTATION?   **91**

ASTRONOMY (CONTINUED)   91

HOW TO MULTIPLY TWO POWERS OF 10?   **93**

ASTRONOMY (CONTINUED)   93

HOW TO DIVIDE BY A POWER OF 10?   **96**

ASTRONOMY (CONTINUED)   96

HOW TO COMPARE POWERS OF 10?   **102**

ASTRONOMY (CONTINUED)   102

YOUR TURN!   **104**

DECIMAL NUMBERS   **105**

EXERCISE 1   105

EXERCISE 2   105

EXERCISE 3   105

EXERCISE 4   105

EXERCISE 5   105

EXERCISE 6   106

EXERCISE 7   106

RELATIVE NUMBERS   **107**

EXERCISE 8   107

EXERCISE 9   107

EXERCISE 10   107

EXERCISE 11   107

FRACTIONS	**108**
EXERCISE 12	108
EXERCISE 13	108
EXERCISE 15	108
EXERCISE 16	109
EXERCISE 17	109
EXERCISE 18	109
EXERCISE 19	109
SQUARES AND SQUARE ROOTS	**110**
EXERCISE 20	110
EXERCISE 21	110
POWERS	**111**
EXERCISE 22	111
EXERCISE 23	111
EXERCISE 24	111
EXERCISE 25	112
SOLUTIONS	**113**
EXERCISE 1	113
EXERCISE 2	114
EXERCISE 3	115
EXERCISE 4	118
EXERCISE 5	119
EXERCISE 6	120
EXERCISE 7	123
EXERCISE 8	124
EXERCISE 9	125
EXERCISE 10	126
EXERCISE 11	127
EXERCISE 12	127

EXERCISE 13   128

EXERCISE 14   129

EXERCISE 15   130

EXERCISE 16   132

EXERCISE 17   132

EXERCISE 18   133

EXERCISE 19   133

EXERCISE 20   134

EXERCISE 22   134

EXERCISE 22   135

EXERCISE 23   136

EXERCISE 24   136

EXERCISE 25   136

CONCLUSION   **139**

Introduction

How many times in your life have you been in a situation where you said to yourself, "If only we could explain it simply! "? Either when you were a student and you had faced a bad teacher, either at a meeting in which the speaker could not reach the level of his audience or when you followed an interview on television and did not understand what the interviewee was saying. So what is your reaction? You pick, you don't listen and in the end you do not learn. I have personally met all these situations and therefore I try to always make the simplest things possible when I have to talk about a subject. Mathematics is omnipresent in everyday life, every day everyone is confronted by numbers: reading the newspapers, watching the news, while shopping, following a recipe, tinkering etc.

Mathematics is not only for clever people. Many people find themselves in a cold sweat at the mention of the word "mathematics". Generations of people are traumatized by how they were taught math. Many feel at a loss in their daily life when they need to calculate quickly. Through this book, I want to reconcile all these people with math and show them how useful mathematical concepts in everyday life can be addressed easily. This book is therefore aimed at an audience of 7-77 years as we used to say. The young student, who learns math in class, will find an original explanation; adults will find items they wanted to have learned at school and they will use in their daily life; older people will have the opportunity to refresh their memory and submit their brains to healthy gymnastics. It has also been proven that regularly exercising the brain is beneficial to the maintenance of memory.

Rarely does a book teach mathematical concepts with a simple approach. In this first volume, limited to 100 pages to address only the basics in an acceptable time reading, we will focus on decimal numbers, relative numbers, squares and square roots, fractions and powers of numbers through very concrete examples. At the end of the book, you can test your knowledge through 25 exercises covering all the concepts studied throughout the book. I hope every reader closes the book with the satisfaction of having learned something useful and having a lot of fun. If this is the case, you will be engaged on the path to healing your allergy to maths!

Integers and decimals numbers

Every day, whether by shopping at the supermarket, watching TV or reading your newspaper, you are confronted with numbers. Basically, all these numbers can be classified into two categories: integers and decimals.

An integer is a number in which there is no decimal point :
4; 9; 24 or 1547 are integers since they do not contain a decimal point.

A decimal number is a number in which there is a decimal point :
4.54; 9.856; 24.006 or 1547.8 are decimal numbers because they contain a decimal point.

Sort decimal numbers

In this chapter, we will learn to manipulate integers and decimals and compare them through simple and concrete examples.

American volume units

When traveling to the United States, you have to adapt to a measurement system completely different from the one used in Europe: the lengths and volumes are expressed in different units as illustrated below:

American Unit	European Unit
Quart (qt)	0.946 L
Cubic foot	28.320 L
Fluid ounce (fl oz)	0.0296 L
Pint (pt)	0.473 L
Gallon (US gallon, gal)	3.785 L

The fluid ounce (fl oz) will be used to buy almost all liquids in supermarkets, a gallon will meanwhile used at service stations to buy fuel.

Exercise: place the different units in the increasing order of their capacity.

Sort numbers in ascending order to rank them from the smallest to the greatest.

The difficulty here is related to the presence of decimals.

To classify decimal numbers up, always start by watching the number that appears before the decimal point: in the above table, the numbers before the decimal point are in the order 0 for the quarter, 28 for cubic foot, 0 for the fluid ounce, 0 for a pint and 3 for a gallon.

This first step gives a first classification: we cannot decide between a quarter ounce and liquid pint but we can see that the cubic feet and the gallon are the greatest.

We have now to consider the quart, the fluid ounce and the pint. To do this, observe the first digit after the decimal point: in the above table, the first decimal is 9 for the quarter, 0 for fluid ounce and 4 for a pint.

This second step will announce that fluid ounce is the smallest, followed by the pint, followed by a quarter.

Combining the results obtained in the previous steps, we can establish the following growing rank:

American Unit	European Unit
Fluid ounce (fl oz)	0.0296 L
Pint (pt)	0.473 L
Quart (qt)	0.946 L
Gallon (US gallon, gal)	3.785 L
Cubic foot	28.320 L

To compare decimals:

Start by comparing the numbers before the decimal point;

If the numbers before the decimal point are identical, then compare the first digit after the decimal point;

If these first figures are identical, then compare the second decimal place;

If these figures are identical 2nd, then compare the 3rd decimal place, and so on ...

A little astronomy

The table below shows the diameters (in thousands of kilometers) of the planets of the solar system:

Planet	Diameter
Earth	12.7
Uranus	50.7
Venus	12.1
Jupiter	143
Mercury	4.9
Mars	6.8
Pluto	2.3
Neptune	49.2
Saturn	120.5

Exercise: Arrange the planets from the largest to the smallest.

As before, we start to arrange the planets in relation to the number located before the decimal point. We realize that no planets have the same number before the decimal point, with the exception of Venus and Earth (12). One can classify all the planets according to their number before the decimal point except Venus and Earth. For these two planets, then observe the first digit after the decimal point. It is 7 for Earth and 1 for Venus; we deduce that Earth is bigger than Venus so we obtain:

Planet	Diameter
Jupiter	143
Saturn	120.5
Uranus	50.7
Neptune	49.2
Earth	12.7
Venus	12.1
Mars	6.8
Mercury	4.9
Pluto	2.3

The Olympic Games

At the races at the Olympic Games, the winners are designated according to the time they take to travel a given distance. The differences between the competitors are generally very low and it is therefore necessary to observe the decimal number to decide between them. Here is the result of the final of the Men's 100 meters at the London Olympics 2012:

Athlete	Time (seconds)
Richard Thompson	9.98
Asafa Powell	11.99
Tyson Gay	9.80
Yohan Blake	9.75
Justin Gatlin	9.79
Usain Bolt	9.63
Ryan Bailey	9.88
Churandy Martina	9.94

Exercise: Rank the athletes from fastest to slowest.

To start, here's a concept vocabulary. When we watch the races, whether in athletics or swimming, it is often necessary to use decimal numbers because the athletes' times are often very close.

When we watch a time in seconds:
The first decimal place is called 10th second $(1/10^{th} = 0.1)$
The second decimal place is called 100th of a second $(1/100^{th} = 0.01)$
The third decimal place is called 1000th of a second $(1/1000^{th} = 0.001)$.

As before, we first rank athletes with the number before the decimal point. Here all have the same number (9) except Asafa Powell (11). We can already deduce that Asafa Powell took the most time and has finished last in the race.

Now compare the 10th second of the athletes with the same number before the decimal point:
An athlete has a 6, two athletes have a 7, two athletes have an 8 and two athletes have a 9. It can be inferred that the athlete who has a 6, Usain Bolt is the fastest and the winner of the race.

Now look at the second and third in the race (those who have a 7 in 1 / 10th of a second) by observing the 100th second: Yohan Blake has a 5 then Justin Gatlin has a 9. It follows that Yohan Blake is the second in the race and Justin Gatlin finished third in the race.

Now look at the fourth and fifth in the race (those with an 8 1 / 10th of a second) by observing the 100th second: Tyson Gay has a 0 while Ryan Bailey has 8. It follows that fourth in the race is Tyson Gay and Ryan Bailey finished fifth in the race.

To finish, look at the sixth and seventh in the race (those who have a 9 in 1 / 10th of a second) by observing the 100th second: Churandy Martina has a 4 while Richard Thompson has 8. It follows that sixth in the race is Churandy Martina, and Richard Thompson finished seventh in the race.

This gives the final standings:

Athlete	Time (seconds)
Usain Bolt	9.63
Yohan Blake	9.75
Justin Gatlin	9.79
Tyson Gay	9.80
Ryan Bailey	9.88
Churandy Martina	9.94
Richard Thompson	9.98
Asafa Powell	11.99

Calculate decimals

In everyday life, most calculations on decimals may be made by a calculator; however there are some techniques for you to make operations more simply and sometimes mentally.

To the market

You go to the market to buy some fruit and vegetables. Your basket weighs 426 g and you add to it:

Fruits / Vegetables	Weight
Cherries	0.7 kg
White grapes	1.75 kg
Carrots	2.478 kg
Tomatoes	4.8 kg

Exercise: calculate the weight of the purchased fruits and vegetables.

To begin, you must ensure that the weight of each item is expressed in the same unit. Sometimes the weight of some items will be expressed in grams (g), while for others it will be expressed in kilograms (kg). In this case, before making the calculation, all the weights should be expressed in the same units (i.e. kg, g). Here all the weights are in kilograms (kg), so the unit is the same.

To calculate the total weight of the fruit and vegetables, you have to add the weight of the cherries, grapes, carrots and tomatoes: 0.7 + 1.75 + 2.478 + 4.8 (kg).

To be sure not to go wrong in the calculation, it is necessary that all numbers have the same number of digits after the decimal point. In our example, the weight of carrots contains more decimal places; there are 3 digits. So, all other weights should be written with 3 figures after the decimal point.

How to do this?

Just add 0 until 3 decimal places appear, which gives:

Fruits / Vegetables	Weight
Cherries	0.7**00** kg
White grapes	1.75**0** kg
Carrots	2.478 kg
Tomatoes	4.8**00** kg

One can now perform the calculation by setting the addition:

```
      0.700
+     1.750
+     2.478
+     4.800
  _____
=     9.728
```

The total weight of fruits and vegetables purchased is 9.728 kg.

To add decimals, make sure that all the numbers have the same amount of digits after the decimal point. If this is not the case, add 0 in order make the decimal numbers have the same amount of digits after the decimal point.

Exercise: calculate the weight of the basket full of fruits and vegetables purchased.

To answer this question we have to add the weight of the basket (426 g) and the weight of fruits and vegetables bought (9.728 kg). We need to do a conversion. We will see conversion in the following chapter.

How to convert in different units?

To the market (continued)

As seen above, you must ensure that the weight of each item is expressed in the same unit. Our basket weighs 426 g and our purchases weigh 9.728 kg. We find that the units are different. So we need to specify if the weight is in grams (g) or kilograms (kg). Here is the table of weights conversion:

kg	hg	dag	g	dg	cg	mg
kilograms	hectograms	decagrams	grams	decigrams	centigrams	milligrams

From this table, it will be very easy to transform grams in milligrams or kilograms etc...

In our example, we will convert the weight of the basket expressed in grams, to kilograms, which is the weight unit of our purchases.

To do this, we put 426 grams in the table, taking care to place the 6 in the grams box and add a decimal point:

kg	hg	dag	g	dg	cg	mg
	4	2	6.			

We would convert these 426 grams to kilograms, which is why in the table, we add 0 as necessary until you reach the column kilograms and we move the decimal point:

kg	hg	dag	g	dg	cg	mg
0.	4	2	6			

It follows that 426 g = 0.426 kg and therefore we can now set the addition:

$$
\begin{aligned}
& \quad \ 0.426 \\
+ \ & \quad 9.728 \\
\hline
= \ & \ 10.154
\end{aligned}
$$

It follows that our basket of fruit weighs 10.154 kg.

Exercise: Convert the total weight of the basket to hectograms then to milligrams.

To do this, we place 10.154 kg in the table, taking care to place a number in each box and place the last digit before the decimal point in the box kilograms:

	kg	hg	dag	g	dg	cg	mg
1	0.	1	5	4			

We would convert these 10.154 kg to hectograms, which is why in the table; we move the decimal point in the column of hectograms:

	kg	hg	dag	g	dg	cg	mg
1	0	1.	5	4			

It follows that 10.154 kg = 101.54 hg

We would convert these 10.154 kg to milligrams, so in the table, we add 0 as necessary to reach the milligrams column then we move the decimal point into the same column:

	kg	hg	dag	g	dg	cg	mg
1	0	1	5	4	0	0	0.

It follows that 10.154 kg = 10154000 mg

Before performing an operation on decimal numbers, make sure that they are expressed in the same unit.
If this is not the case, make the conversion using the conversion table.

How to subtract decimals?

To the market (continued)

Suppose now that we remove the tomatoes from our basket.

Exercise: Calculate the total weight of the basket without tomatoes.

From the total weight of the basket previously determined (10.154 kg), we must subtract the weight of tomatoes (4.8 kg).

To do this, we check that the weights are expressed in the same unit: this is the case here since the two weights are in kilograms;

Before setting subtraction, we must ensure that both numbers have the same amount of digits after the decimal point: this is not the case here. So we have to write 4.8 kg as follows: 4.800 kg.

We can now set the subtraction:

$$
\begin{array}{r}
10.154 \\
-\quad 4.800 \\
\hline
=\quad 5.354
\end{array}
$$

After removing the tomatoes, the basket now weighs only 5.354 kg.

To subtract decimals, make sure that the numbers have the same amount of digits after the decimal point. If this is not the case, add some zeros in order to set the subtraction with decimal numbers having the same amount of digits after the decimal point.

How to multiply decimals?

To the market (continued)

Cauliflower is on promotion in this market since it is offered at € 1.65 per kilo. You decide to buy, and your bag contains 1.583 kg.

Exercise: Calculate the price of cauliflower.

You know that every kilo of cauliflower costs € 1.65.
You buy 1.583 kg.

To get the total price, just make the following multiplication: 1.583 x 1.65.

You multiply two decimal numbers; for this calculation it is easier at first to ignore the decimal points and set the following multiplication:

$$
\begin{array}{r}
1583 \\
x \quad 165 \\
\hline
= \quad 261195
\end{array}
$$

In a second step, it is interesting to see that 1.583 has 3 decimal places, and 1.65 has 2 decimal places: this means that the result of 1.583 x 1.65 has 3 + 2 = 5 digits after the decimal point.

In the multiplication result obtained previously, just count now 5 digits from the end and put the decimal point before the fifth digit: 2.61195.

This means that 1.583 x 1.65 = 2.61195.

Rounding the result to two decimal places, we deduce that the cauliflower cost us € 2.61.

To multiply two decimal numbers together:

Perform the multiplication regardless of decimal points.
Count the amount of decimal places in each of the numbers.
Add up the amount of digits, this gives you the amount of decimal places in the result.

How to divide decimals?

In the last operation we have not addressed division. Division is the process that you need to achieve when you want to share an item (a pie, candy, money) fairly among several people.

Suppose you have a package containing 12 cookies and want to share it equitably between your 3 children. You then divide 12 by 3 to give 4. This means that you spread your 12 biscuits into 3 lots each containing 4 cookies and therefore each of your 3 children will receive four biscuits.
Once you have made the division, you may also find that no biscuits remain. This is because the result of the division won't have any remainder, then we say that 12 can be divided by 3. We can write 12 = 3 x 4 + 0.

Vocabulary:

12 is called the dividend (the number that you want to share)
3 is called the divisor (the number of units you want to allocate)
4 is called the quotient (the result of the division: what's in each of the shares)
0 is called the remainder (what's left of the number to share when the shares were distributed).

Now suppose you have 25 candies to share equally among your four nephews. As before, you will divide 25 by 4. If you make this calculation by head or hand, you will find 6 with a remainder of 1. This means that each of your four nephews receives 6 sweets and that there will remain 1 undistributed candy (you cannot share one candy among 4 persons!). If you perform this calculation on the calculator, you will find 6.25.

As the result of your division, this leaves a remainder (here: 1 candy), or gives a decimal to the calculation which means 25 cannot be divided by 4. We can write $25 = 4 \times 6 + 1$.

The division is the process that allows for sharing.

When you are dividing a number A by a number B:

If the hand calculation shows a remainder not equal to 0, this means that the number A is not divisible by the number B.
If the calculation in the calculator indicates a decimal number, this means that the number A is not divisible by the number B.

If the hand calculation shows a remainder equal to 0, this means that the number A is divisible by the number B,
If the calculation in the calculator shows an integer, it means that the number A is divisible by the number B.

There are tricks that allow you to easily find out if a number is divisible by another number between 1 and 13. This trick is called "divisibility criteria" and was explained in my book "The secrets of mental math, everyone is able to calculate in a flash!"

Divisibility criteria	Example
Divisibility by 2: A number is divisible by 2 if its last digit is even (it finishes by 0, 2, 4, 6 or 8)	126 is divisible by 2. 133 is not divisible by 2.
Divisibility by 3: A number is divisible by 3 if the addition of its digits is a multiple of 3.	131 is not divisible by 3 because 1+3+1=5, and 5 is not a multiple of 3. 531 is divisible by 3 because 5+3+1=9, and 9 is a multiple of 3.
Divisibility by 4: A number is divisible by 4 if its two last digits are a multiple of 4.	311 is not divisible by 4 because 11 is not a multiple of 4. 624 is divisible by 4 because 24 is a multiple of 4.

Divisibility by 5: A number is divisible by 5 if its last digit is a 0 or a 5.	234 is not divisible by 5 because its last digit is 4. 990 is divisible by 5 because its last digit is 0.
Divisibility by 6: A number is divisible by 6 if it is at the same time divisible by 2 and by 3.	741 is not divisible by 6 because it is not divisible by 2 (but it is divisible by 3). 234 is divisible by 6 because it is at the same time divisible by 2 and by 3.
Divisibility by 7: A number is divisible by 7 if ht − (ux2) is divisible by 7. In the number: h is the digit of hundreds, t the digit of tens and u the digit of units.	176 is not divisible by 7 because $17 - 2 \times 6 = 17-12 = 5$ is not divisible by 7. 553 is divisible by 7 because $55 - 2 \times 3 = 55-6 = 49$ is divisible by 7.

Divisibility by 8:	
A number is divisible by 8 if ht + (u/2) is divisible by 4. In the number: h is the digit of hundreds, t the digit of tens and u the digit of units.	834 is not divisible by 8 because ht + (u/2) = 83 + (4/2) = 85 is not divisible by 4. 616 is divisible by 8 because ht + (u/2) = 61 + (6/2) = 64 is divisible by 4.
Divisibility by 9: A number is divisible by 9 if its digit addition is a multiple of 9.	445 is not divisible by 9 because 4+4+5=13 and 13 is not divisible by 9, 756 is divisible by 9 because 7+5+6=18 and 18 is divisible by 9.
Divisibility by 10: A number is divisible by 10 if its last digit is 0.	849 is not divisible by 10 because its last digit is 9. 320 is divisible by 10 because its last digit is 0.

Divisibility by 11: A number is divisible by 11 if the difference between the even digit addition and the odd digit addition is divisible by 11.	13<u>5</u>4 is not divisible by 11 because <u>1</u>+<u>5</u>=6 and **3+4**=7 and 7-6=1 is not divisible by 11. 13<u>6</u>4 is divisible by 11 because <u>1</u>+<u>6</u>=7 and **3+4**=7 and 7-7=0 is divisible by 11.
Divisibility by 12: A number is divisible by 12 if it is at the same time divisible by 3 and by 4.	525 is not divisible by 12 because it is not divisible by 4 (but it is divisible by 3). 156 is divisible by 12 because it is at the same time divisible by 3 and by 4.
Divisibility by 13: A number is divisible by 13 if the addition of the number of tens and the quadruple of the unit digits is divisible by 13.	426 is not divisible by 13 because 42 + 4x6 = 42+24=68 is not divisible by 13. 637 is divisible by 13 because 63 + 4x7 = 63+28=91 is divisible by 13. 91 is divisible by 13 because 9 + 4x1 = 9+4 = 13 is divisible by 13.

The move

You prepare your move and feel that you will need 175 boxes to pack all your things. The supply store sells only those boxes in batches, each containing 14 boxes.

Exercise: Calculate how many batches of 14 boxes are needed for your move and then determine if you have any remaining unused boxes.

To start, you have to determine what you want to share and then how many parts you must share it into. In our example, you know that you need a total of 175 boxes; it is the quantity to share. These boxes are only sold in batches, i.e. in units each containing 14 boxes. To answer the question, you have to determine how many batches, each containing 14 boxes, are required to obtain 175 boxes. To do this, 175 is divided by 14.

By performing the division by hand, you will find 14 x 175 = 12 + 7. By calculator you can find 12.5. This means that to have 175 boxes, you need 12 batches of 14 cartons and 7 other boxes.

Unfortunately boxes are not sold in units, that is why you need to buy 13 batches of 14 boxes, which are 13 x 14 = 182 boxes. You have 7 boxes more than necessary (for 182 - 175 = 7).

Finally, to ensure your move you have to buy 13 batches of 14 boxes and you are left with 7 unused boxes.

The stamp collector

A stamp collector proudly announces: "I have more than 400 but fewer than 450 stamps!
If I group by 2, by 3, 4 or 5, I still have one stamp left!"

Exercise: Determine how many stamps the collector has.

To solve this problem, it is necessary to use the criteria of divisibility.
The collector said that if he groups his stamps by 5, it remains to 1. This means that the amount of stamps is not divisible by 5. Now, according to the table of divisibility criteria, a number can be divided by 5 if it ends in 0 or 5. As there always remains one stamp by grouping the stamps by 5, we deduce that this collector has either 401, 406, 411, 416, 421, 426, 431, 436 441 or 446 stamps.

Similarly, the collector says that if he groups stamps by 2, he still has one.
This means that the amount of stamps is not divisible by 2. However, according to the table of divisibility criteria, a number is divisible by 2 if it is an even number. We deduce that this collector has an odd number of stamps. Among the possibilities enumerated above, we can remove even numbers; it leaves 401, 411, 421, 431 and 441 stamps.

The collector said that if he groups stamps by 4, he still has one.

This means that the amount of stamps is not divisible by 4. However, according to the table of divisibility criteria, a number is divisible by 4 if the number formed by the last two digits is divisible by 4.

Among the possibilities enumerated above:
401 is possible because 400 is divisible by four (00 is divisible by 4), and remains 1
411 is not possible because 410 is not divisible by 4 (10 is not divisible by 4)
421 is possible because 420 is divisible by 4 (20 is divisible by 4), and remains 1
431 is not possible because 430 is not divisible by 4 (30 is not divisible by 4)
441 is possible because 440 is divisible by 4 (40 is divisible by 4), and remains 1.
At this stage of our investigation, the collector has either 401, 421 or 441 stamps.

Now study the divisibility by 3.
The collector said that if he groups stamps by 3, he still has one.
This means that the amount of stamps is not divisible by 3. However, according to the table of divisibility criteria, a number is divisible by 3 if the sum of its digits is divisible by 3.
401 is not divisible by 3 because 4 + 0 + 1 = 5 and 5 is not divisible by 3. However, 401 = 133 x 3 + 2. The rest is not 1, so 401 cannot be the answer.
421 is not divisible by 3 because 4 + 2 + 1 = 7 and 7 is not divisible by 3. However, 421 = 140 x 3 + 1. The remainder is 1, so 421 may correspond to the number of the stamp collection.

36

441 is divisible by 3 because 4 + 4 + 1 = 9. So 441 cannot match the number of collector's stamps since it is not supposed to be divisible by 3 (1 stamp remains).

We can conclude with certainty that our collector has exactly 421 stamps.

Good accounts make good friends

A group of fewer than 40 people must equitably distribute a sum of € 229. This leaves € 19. Another time, the same group must fairly distribute € 474: this time, there is € 12 left.

Exercise: Determine how many people are in this group.

Call P the number of people in the group. When seeking to share € 229 between the P people, the final € 19 remains. This means that 229 is not divisible by P, however 229-19 = 210 is divisible by P.

Similarly, when trying to share € 474 between the P people, there is € 12 left. This means that 474 is not divisible by P, however 474-12 = 462 is divisible by P.

So we have to find all the numbers that are divisors of both 210 and 462.

To find all the divisors of these numbers, the key is to break up the numbers 210 and 462 as follows:

210 = 21 x 10 = 3 x 7 x 5 x 2 (we cannot break 210 with smaller numbers than 2, 3, 5 and 7).

462 = 231 = 2 x 2 x 3 x 77 = 2 x 3 x 7 x 11 (we cannot divide 462 with smaller numbers than 2, 3, 7 and 11).

From the expression 210 = 3 x 7 x 5 x 2, we can write (by grouping the numbers 3, 7, 5 and 2 in different ways):

210 = 3 x (7 x 5 x 2) = 3 x 70
210 = 7 x (3 x 5 x 2) = 7 x 30
210 = 5 x (3 x 7 x 2) = 5 x 42
210 = 2 x (3 x 7 x 5) = 2 x 105
210 = (3 x 7) x (5 x 2) = 21 x 10
210 = (3 x 2) x (7 x 5) = 6 x 35
210 = (3 x 5) x (7 x 2) = 15 x 14

From this, we can guess that 2, 3, 5, 6, 7, 10, 14, 15, 21, 30, 35, 42, 70 and 105 are divisors of 210.

Similarly, from the expression 462 = 2 x 3 x 7 x 11, we can write:

462 = 2 x (3 x 7 x 11) = 2 x 231
462 = 3 x (2 x 7 x11) = 3 x 154
462 = 7 x (2 x 3 x 11) = 7 x 66
462 = 11 x (2 x 3 x 7) = 11 x 42
462 = (2 x 3) x (7 x 11) = 6 x 77
462 = (3 x 7) x (2 x 11) = 21 x 22
462 = (2 x 7) x (3 x 11) = 14 x 33

From this, we can guess that 2, 3, 6, 7, 11, 14, 21, 22, 33, 42, 66, 77, 154 and 231 are divisors of 462.

We can see that the numbers which are both dividers of 210 and 462 (which are found in the two expressions) are 2, 3, 6, 7, 14, 21 and 42. Among these numbers is the number of P people in the group.

We know that the group has fewer than 40 people, so we can eliminate 42.

If the group contained 2 persons, the remainder of 229 by 2 would be 1 and not 19 (as 229 = 2 x 114 + 1): we can eliminate 2,

If the group contained 3 persons, the remainder of 229 by 3 would be 1 and not 19 (as 229 = 3 x 76 + 1): we can eliminate 3,

If the group contained 6 persons, the remainder of 229 by 6 would be 1 and not 19 (as 229 = 6 x 38 + 1): we can eliminate 6,

If the group contained 7 persons, the remainder of 229 by 7 would be 5 and not 19 (as 229 = 7 x 32 + 5): we can eliminate 7,

If the group contained 14 persons, the remainder of 229 by 14 would be 5 and not 19 (as 229 = 14 x 16 + 5): we can eliminate 14,

If the group contained 21 persons, the remainder of 229 by 21 would be 19 (as 229 = 21 x 10 + 19): we can guess that the group contains 21 persons.

Similarly 21 x 474 = 22 + 12. So by sharing € 474 between 21 people, € 12 remains.

So the group contains 21 people.

The difficult choice of tiles

A rectangular pool measures 3.36 m by 7.80 m by 1.44 m. We want to install identical square tiles. The craftsman does not want to cut tiles and prefers large tiles because they are easier to install. The supplier has all sizes of tiles (with integer number of centimeters).

Exercise: Determine the size of the tiles to order.

To start, the tile size is expressed in centimeters while the pool dimensions are in meters. We have to harmonize units. Generally, it is best to choose the smallest unit: here we choose the centimeter. Thus the pool measures 336 cm by 780 cm by 144 cm.

The tiles we choose must be square.

For an integer number of tiles in a width of 336 cm, 336 must be divisible by the dimensions of a tile. Indeed, if 336 is not divisible by the size of a tile, this means that the division has a remainder and we have to cut a tile.

As in the previous exercise, we have to decompose 336.

We can write that 336 = 3 x 112 = 3 x 4 x 28 = 3 x 4 x 4 x 7 = 2 x 2 x 2 x 2 x 3 x 7

Making groups as in the previous exercise:

336 = 2 x 168
336 = 3 x 112
336 = 7 x 48

41

336 = 4 x 84
336 = 6 x 56
336 = 8 x 42
336 = 14 x 24
336 = 12 x 28
336 = 16 x 21

Dividers of 336 are 2, 3, 4, 6, 7, 8, 12, 14, 16, 21, 24, 28, 42, 48, 56, 84, 112 and 168. This means the width of the pool may be tiled with an integer number of tiles, when these ones have a width equal to one of the above numbers.

Similarly, for an integer number of tiles on a length of 780 cm, 780 must be divisible by the dimension of a tile. Indeed, if 780 is not divisible by the size of a tile, it would mean that the division has a remainder and we have to cut a tile.

We are looking for dividers of 780, considering that: 780 = 10 x 78 = 5 x 2 x 2 x 39 = 5 x 2 x 2 x 3 x 13

So 780 = 2 x 390
780 = 3 x 260
780 = 5 x 156
780 = 13 x 60
780 = 4 x 195
780 = 6 x 130
780 = 10 x 78
780 = 12 x 65
780 = 20 x 39
780 = 26 x 30
780 = 15 x 52

Dividers of 780 are 2, 3, 4, 5, 6, 10, 12, 13, 15, 20, 26, 30, 39, 52, 60, 65, 78, 130, 156, 195, 260 and 390. This means that length of the pool may be tiled with an integer number of tiles, when these ones have a length equal to one of the above numbers.

Finally, to have an integer number of tiles for a height of 144 cm, 144 must be divisible by the size of a tile. Indeed, if 144 is not divisible by the size of a tile, it would mean that the division has a remainder and we have to cut a tile.

We search the dividers of 144, considering that: $144 = 48 = 3 \times 3 \times 4 \times 12 = 3 \times 2 \times 2 \times 2 \times 2 \times 3$

So $144 = 2 \times 72$
$144 = 3 \times 48$
$144 = 4 \times 36$
$144 = 6 \times 24$
$144 = 8 \times 18$
$144 = 9 \times 16$
$144 = 12 \times 12$

The dividers of 144 are 2, 3, 4, 6, 8, 9, 12, 16, 18, 24, 36, 48 and 72. This means that the height of the pool can be tiled with an integer number of tiles, if these ones have a height equal to one of the above numbers.

As the tiles are square, they have the same size in width and length. We have to retain numbers that are both dividers of 336, 780 and 144. In the above lists, only the numbers 2, 3, 4, 6, and 12 can correspond (they are present in the 3 lists).

In the statement, it is clear that the craftsman wants large tiles: the best choice is to use square tiles of 12 x 12 x 12 cm.

Fractions

If there is a mathematical subject that is often cited as one that triggered a math phobia to generations of children, it is undoubtedly fractions. It is an esoteric subject: we are told that the fractions are perfectly adapted to the problems of everyday life when cutting a slice of cake, we are told that there is a numerator, denominator, and that we have to bring them to the same denominator to perform calculations ... except that in reality, the concept is complex for everyone. The following chapter aims to make lasting peace with fractions and show you that everyday life is full of opportunities to use fractions, sometimes without even knowing it.

What is a fraction?

The king of cocktails

Suppose you are planning a party with friends and, for the occasion, you decide to prepare a famous cocktail called Bloody Mary with the following recipe:

8 volumes of Vodka,
24 volumes of tomato juice,
1 volume of lemon juice,
1 volume of Worcestershire sauce,
A few drops of Tabasco,
Celery salt,
Pepper

Exercise: Determine the amounts of ingredients used to prepare 3 liters of this cocktail.

This recipe contains 4 fluids which are vodka, tomato juice, lemon juice and Worcestershire sauce. The other ingredients (Tabasco, salt and pepper) are to correct the seasoning. If we sum the volumes of liquid, given in the recipe, we have 8 (vodka) + 24 (tomato) + 1 (lemon) + 1 (sauce) or 34 volumes.

This means that in 34 liters of cocktail, there would be 8 liters of vodka, 24 liters of tomato juice, 1 liter of lemon juice, 1 liter of Worcestershire sauce.

In other words, this means that a given amount Cocktail contains:

46

8 / 34 of vodka,
24 / 34 of tomato juice,
1 / 34 of lemon juice,
1 / 34 of Worcestershire sauce

These quantities are called fractions.

A fraction can be written as A / B where A is called the numerator and the denominator B.

A fraction corresponds to a division in which the number A is divided by the number B.

In our example, this means that whatever the cocktail volume that is taken, it can be "cut" into 34 units. Of these 34 units, 8 will be vodka, 24 will be tomato juice, 1 will be lemon juice and 1 will be Worcestershire sauce.

We say that this cocktail contains 8 34th of vodka, 24 34th of tomato juice, 1 34th of lemon juice and 1 34th of Worcestershire sauce.

Now we need to determine the quantities of each product to prepare 3 L of cocktail.

Our 3L of cocktail can be cut into 34 units, that means 3/34 "batches" of cocktail.
For vodka, each batch (3/34) contains 8 units of vodka i.e. 8 X 3/34 = 24/34 vodka,

For tomato juice, each batch (3/34) contains 24 units of tomatoes i.e. 24 x 3/34 = 72/34 tomatoes,

For lemon juice, each batch (3/34) contains 1 unit of lemon i.e. 1 x 3/24 = 3/34 lemon,

For Worcestershire sauce, each batch (3/34) contains 1 unit of sauce i.e. 1 x 3/34 = 3/34 sauce.

The figures show that when the ingredients are put in the form of a fraction, just multiply this fraction by the volume of cocktail that is to be prepared in order to obtain the amount to use for each ingredient.

To prepare 3 L Bloody Mary, we need:
24/34 = 0.7 L of vodka,
72/34 = 2.1 L of tomato juice,
3/34 = 0.09 L of lemon juice,
3/34 = 0.09 L of Worcestershire sauce

With the conversion table below it is possible to express the amounts of liquid in liters (L) into a more easily measurable unit (cL or mL for example):

	L	dL	cL	mL
Vodka	0.	7	0	0
tomato juice	2.	1	0	0
lemon juice	0.	0	9	0
Worcestershire sauce	0.	0	9	0

To prepare 3 L of Bloody Mary, we need:

0.7 L = 70 cL = 700 ml of vodka,
2.1 L = 210 cL = 2100 ml of tomato juice,
0.09 L = 9 cL = 90 ml of lemon juice,
0.09 L = 9 cL = 90 mL of Worcestershire sauce.

Calculate with fractions

House for sale

An apartment of 86 sqm is offered for sale at a price of €
162,000.

Exercise: Determine the price of square meters of this apartment.

This apartment has an area of 86 m²; it means it can be cut into 86 small pieces of 1m² each. To solve our problem we need to find the price of one of these little pieces of 1 m².

To do this, we divide the price of the apartment (€ 162,000) by 86 small pieces of 1 m² which gives 162,000 / 86.

We have here a fraction whose numerator is 162,000 and the denominator is 86. The calculation of this fraction gives 1884 (rounded).

This apartment is proposed at a price of € 1,884 per sqm.

We take credit?

A car dealer offers a vehicle on favorable terms: the customer pays a third of the vehicle on the day of purchase and the remainder in 24 monthly payments at no charge. The vehicle costs € 20,520.

Exercise: Determine the amount to be paid on the day of purchase and the amount of each monthly payment.

The customer has to pay a third of the vehicle on the day of purchase which is a fraction of 1/3.

This also means that he will have to pay two thirds of the vehicle over the next 24 monthly payments.

The vehicle price is € 20,520.

> **A fraction of a quantity means to multiply the quantity by the fraction.**
>
> **To multiply a fraction A / B by a number C :**
>
> **First calculate A x C then divide the result by B : (A x C) / B**

A third of the vehicle price is 20520 x (1/3) = (20520 x 1) / 3 = 20520/3 = 6840.

2/3 of the vehicle price is 20520 x (2/3) = (20520 x 2) / 3 = 41040/3 = 13680.

This means that the customer will pay € 6,840 on the day of purchase and € 13,680 in 24 monthly payments.

We have now to determine how much the customer will pay in each monthly payment.

To do this, we divide the price still to be paid (13,680) by the number of payments (24) which gives the following fraction 13680 / 24. The calculation of this fraction gives 570.

We have all the answers to the problem: the customer will pay € 6,840 on the date of purchase of the vehicle and € 570 per month for 24 months.

This is a survey

252 employees were questioned about their lunch break.

1/6 of the employees never eat in the canteen,
3/7 of the employees eat only once a week in the canteen,
3/14 employee eat twice a week in the canteen,
The rest of the employees eat more than twice a week in the canteen.

Exercise: Determine how many employees are in each category.

We are told that 1/6 of employees never eat in the canteen. This means that if we share all employees into groups of 6, 1 employee in each group would not eat in the canteen. This leads to the following calculation: 252 x (1/6)

252 x (1 / 6) = 252 / 6 = 42

Similarly we are told that 3/7 of employees eat once a week in the canteen. This means that if we share all employees into groups of 7, 3 employees in each group would eat once a week in the canteen. This leads to the following calculation: 252 x (3/7)

252 x (3 / 7) = (252 x 3) / 7 = 756 / 7 = 108

Similarly we are told that 3/14 employee eats twice a week in the canteen. This means that if we share all

53

employees into groups of 14, 3 employees in each group would eat twice a week in the canteen. This leads to the following calculation: 252 x (3/14)

252 x (3 / 14) = (252 x 3) / 14 = 756 / 14 = 54

We therefore deduce that:

42 employees never eat in the canteen,
108 employees eat once a week in the canteen,
54 employees eat twice a week in the canteen.

All remaining employees (252 - 42 — 108 - 54 = 48) eat more than twice a week in the canteen.

Compare fractions

The legacy of an eccentric uncle

A great uncle wrote his will and decided to give all his property to his nephews. Nevertheless he divided his property into strange parts:

3 / 12   3 / 4   3 / 8   5 / 6   1 / 3

Exercise: Determine who will receive the largest share.

Here we have five fractions to be compared from the smallest to the largest or vice versa. The problem is that, without a calculator, it is rather complicated to know if 3/4 is larger or smaller than 5/6 ... It would be much easier to compare these fractions if they had the same denominator: in fact, if I give you 2/6 and 5/6, it is obvious that 5/6 is larger than 2/6 since we have more by cutting 5 in 6 shares (5/6) than by cutting 2 in 6 shares (2 / 6).

When you wish to compare two fractions, always put the same denominator.

When two fractions have the same denominator, the larger of the two is the one with the largest numerator.

In our example, we have 5 fractions with different denominators: 5; 12; 4; 8; 6 and 3. To give the same denominator to these fractions, we must seek a common number to 12, 4, 8, 6 and 3 ... In other words, what number appears in the multiplication tables of 12, 4, 8, 6 and 3?

24 is the answer because:
12 x 2 = 24,
4 x 6 = 24,
8 x 3 = 24,
6 x 4 = 24,
3 x 8 = 24.

If you want to express 3/12 in a fraction whose denominator is 24: you have to multiply the denominator (12) by 2 (because 12 x 2 = 24). In this case, you have to multiply by 2 the numerator too.

And 3/12 = (3 x 2) / (12 x 2) = 6/24
Dividing 3 by 12 gives the same result as dividing 6 by 24!

A fraction does not change when we multiply numerator A and denominator B by the same nonzero number C.

And A / B = (A x C) / (B x C)

Similarly to express the fraction 3/4 in a fraction whose denominator is 24: I have to multiply top and bottom by 6, where:

3 / 4 = (3 x 6) / (4 x 6) = 18 / 24

Similarly to express the fraction 3/8 in a fraction whose denominator is 24: I have to multiply top and bottom by 3, where:

3 / 8 = (3 x 3) / (8 x 3) = 9 / 24

To express the fraction 5/6 in a fraction whose denominator is 24: I have to multiply top and bottom by 4, where:

5 / 6 = (5 x 4) / (6 x 4) = 20 / 24

To express the fraction 1/3 in a fraction whose denominator is 24: I have to multiply top and bottom by 8, where:

1 / 3 = (1 x 8) / (3 x 8) = 8 / 24

Thus it follows that comparing:

  3 / 12   3 / 4   3 / 8   5 / 6   1 / 3

Returns to compare:

  6 / 24   18 / 24   9 / 24   20 / 24   8 / 24

Now that all these fractions have the same denominator, it is easier to classify these fractions from the largest to the smallest, because you only have to compare their numerators, which gives:

20 / 24   18 / 24   9 / 24   8 / 24   6 / 24

By taking the original fractions, here you have from the largest to smallest:

5 / 6   3 / 4   3 / 8   1 / 3   3 / 12

Simplify fractions

Let's play Scrabble®!

Scrabble® is a popular word game composed of tiles each marked with a letter of the alphabet. The distribution of tiles in the game package is the following:

Letter	E	A	I	N O R S T U	L	D M	B C F G H P V White	J K Q W X Y Z
Amount	15	9	8	6	5	3	2	1

Exercise: Determine the total number of chips in the box.

The trap here is to sum directly all the figures on the "Amount" line because in this case you will get fewer tiles that there actually are. Why?

Because the column containing the letters U and N O R S T indicates 6 tiles. That means there are 6 tiles for each of the 6 letters. There are 6 tiles for the letter N, 6 tiles for the letter O and so on ... So N O R S T and U letters represent 6 x 6 = 36 tiles.

Similarly, the two letters D and M (3 tiles each) account for 2 x 3 = 6 tiles.

In the same way:
8 letters B C F G H P V and White count for 8 x 2 = 16 tiles,
7 letters J K Q W X Y and Z count for 7 x 1 = 7 tiles.

In total we have:
15 + 9 + 8 + 36 + 5 + 6 + 16 + 7 = 102 tiles in a box of Scrabble®.

Exercise: Determine what portion of tiles is marked with the letter V.

There are a total of 102 tiles in the box. Among these 102 tiles, 16 are marked with the letter V. This corresponds to a fraction of 16/102.

Exercise: Determine what portion of tiles is marked by the letter E.

There are a total of 102 tiles in the box. Among these 102 tiles, 15 are identified by the letter E. This corresponds to a fraction of 15/102.

Note that this fraction can be simplified. Indeed, the numerator (15) is a multiple of 3 (because 15 = 3 x 5), and the denominator (102) is also a multiple of 3 (for 102 x 3 = 34).
So we can write that 15/102 = (3 x 5) / (3 x 34).

In such a fraction that contains only multiplications, we can delete the duplicate numbers in the numerator and the denominator (here 3) without it changing the value of the fraction.

This means that 15/102 = 5/34 (you can check this with a calculator).

We say we simplified the fraction 15/102 by writing it in a form in which we cannot find a common number in the numerator and denominator.

Exercise : Can you simplify the fractions 11 / 143 ; 6 / 16 ; 10 / 75 et 9 / 13?

To tell if a fraction can be simplified, you have first to find if it is possible to decompose the numerator and denominator in order to make it appear the same number.

For example, in the fraction 11/143, is it possible to display the same number by decomposing 11 and by decomposing 143? 11 is not divisible by any number other than itself (11 x 11 = 1). If 11/143 can be simplified, the only way to proceed is to decompose 143 to show an 11. Note that 11 x 13 = 143.

So 11 / 143 = (11 x 1) / (11 x 13), we can delete 11 in the numerator and in the denominator to obtain 1 / 13. We can guess that 11 / 143 = 1 / 13 (and 1 / 13 cannot be simplified yet).

If we follow the same reasoning for 6/16, we can write that 6/16 = (2 x 3) / (2 x 8). By removing the 2 on the top and bottom, we obtain 3 / 8. So 6/16 can be simplified into 3/8.

10 / 75 = (2 x 5) / (15 x 5). By removing the 5 on the top and the bottom, we obtain 2 / 15. 10 / 75 can be simplified into 2 / 15.

9 / 13 cannot be simplified because the denominator (13) cannot be divided by any number other than itself. And 9 cannot be decomposed with a multiple of 13.

A **fraction can be simplified** if we can decompose the numerator and the denominator to display a common number.

For example, the fraction A / B can be simplified if we can write it as follows (a x c) / (b x c).

The simplified fraction is a / b.

And we have A / B = a / b.

How to multiply fractions?

The grocer

In one day, a grocer sold 3/4 of his cheeses in the morning and 2/3 of the rest in the afternoon.

Exercise: Determine what portion of cheeses he still has at noon, then what fraction of cheeses he sold in the afternoon?

The grocer sold 3/4 of his cheeses in the morning meaning that if he had 4 cheeses early in the day, he would have sold 3 in the morning. 1 cheese of the 4 would have remained to be sold in the afternoon. So at noon, he still has 1/4 of his cheeses.

In the afternoon, he sold 2/3 of the remaining cheeses, that means 2/3 of 1/4. In the exercise "We take credit?" we have seen that taking a fraction of a quantity consists in multiplying the quantity by the fraction. So 2/3 of 1/4 can be written (2/3) x (1/4).

To **multiply two fractions**, multiply numerators and denominators.

So (A / B) x (C / D) = (A x C) / (B x D).

So (2 / 3) x (1 / 4) = (2 x 1) / (3 x 4) = 2 / 12.

2 / 12 can be simplified because 2 / 12 = (2 x 1) / (2 x 6) = 1 / 6.

We deduce that in the afternoon, the grocer sold 1/6 of his cheeses.

How to add fractions?

The best pastry

A pastry chef has prepared 15 kg of cake batter. In the morning, he uses 3/8 of the mixture for the cakes that will decorate his window. In the afternoon, he uses 5/24 of this mixture to make a cake.

Exercise: Determine what weight of batter he used during the day and what weight of dough he still has in the evening?

In one day, the pastry chef used 3/8 + 5/24 of batter.

To add two fractions, you have first to put them into the same denominator.

Then add the numerators together. The denominator remains unchanged.

And (A / B) + (C / B) = (A + C) / B

So to calculate (3/8) + (5/24), we need to put these two fractions into the same denominator. Noting that 24 = 8 x 3, we can write: 3/8 = (3 x 3) / (8 x 3) = 9/24.

So (3 / 8) + (5 / 24) = (9 / 24) + (5 / 24).

As the fractions have the same denominator, we can now add up very simply by adding the numerators:

$(9 / 24) + (5 / 24) = (9 + 5) / 24 = 14 / 24.$

Finally, we can simplify this fraction noting that $14 / 24 = (7 \times 2) / (12 \times 2) = 7 / 12.$

In one day, the pastry chef has used 7/12 of the mixture prepared in the morning. At night, 5/12 remains.

In the exercise "We take credit?" we have seen that taking a fraction of a quantity consists of multiplying the quantity by the fraction.

In the morning, the pastry chef prepared 15 kg of batter and he used 7/12 during the day, that is to say $15 \times (7/12) = (15 \times 7) / 12 = 105 / 12$. With a calculator, we find 8.75.

In the evening, he still has 5/12 of what he prepared or $15 \times (5/12) = (15 \times 5) / 12 = 75 / 12$. With a calculator, we find 6.25.

It follows that, in a preparation of 15 kg of batter, the pastry chef used 8.75 kg during the day and he still has 6.25 kg in the evening.

A generous friend

Matthew won the lottery. He decided to benefit three of his friends. To John, he gives 1/8 of his earnings, to Luke he gives 1/6 of his earnings; to Marc he gives 1/5 of what he has not distributed and keeps the rest for himself.

Exercise : What portion of his gains go to Matthew?

The purpose of this exercise is to determine the fraction of the gains received by each of his friends.

Matthew distributed 1/8 and 1/6 of his winnings to John and Luke. This represents 1/8 + 1 / 6. We have seen that to add two fractions, they must have the same denominator. Here the common denominator of 6 and 8 is 24 because 24 = 8 x 3 and 24 x 4 = 6.

So 1 / 8 + 1 / 6 = (1 x 3) / (8 x 3) + (1 x 4) / (6 x 4) = 3 / 24 + 4 / 24 = 7 / 24.

Matthew has distributed 7/24 of his winnings to John and Luke. This leaves him 24/24 - 7/24 = 17/24 of his gains.

To Marc, he gives 1/5 of what he did not distribute, it means 1/5 of 17/24 = (1/5) x (17/24) = (1 x 17) / (5 × 24) = 17/120.

We wrote John and Luke's shares in 24th (3/24 and 4/24) but Marc's share is expressed in 120th (17/120). To compare those fractions, we must put the same denominator. Here the common denominator is 120, because 120 = 24 x 5.

John's share is 3/24 = (3 x 5) / (24 x 5) = 15/120

Luke's share is 4/24 = (4 x 5) / (24 x 5) = 20/120

Mark's share is 17/120

So the three friends received 15/120 + 20/120 + 17/120 = 52/120

Matthew has meanwhile received the rest, which is 120/120 - 52/120 = 68/120.

Relative numbers

Whether in your newspaper or on television, weather forecasts show temperatures in different cities. In Europe temperatures are expressed in degrees Celsius (°C). In summer, temperatures are positive, in winter you often see temperatures preceded by a minus sign (-). All temperatures are compared to 0 °C. If they are greater than 0 °C, they are positive; if they are less than 0 °C, they are negative and are therefore preceded by a minus sign (-).

The relative position of a number is indicated with respect to 0:
A number greater than 0 is positive while a number less than 0 is negative.

Classify and calculate the relative numbers

In this chapter, we will learn to manipulate these relative numbers and perform calculations through examples drawn from everyday life.

Weather forecasts

Here are the forecasts of temperatures recorded in a newspaper during the winter:

Town	Temperature morning (°C)	Temperature afternoon (°C)
Paris	-4	2
Ajaccio	6	14
Nancy	-9	-1
Bordeaux	8	10

Exercise: Rank these cities from the coldest to the hottest in the morning.

The lower the temperature is, the colder it is. Thus the coldest cities will be those whose temperatures are negative.

When comparing two negative numbers, the smallest is the one that is farthest from 0

The coldest cities are Paris and Nancy (negative temperatures), but -9 is further from 0 than -4, so Nancy is the coldest city.

The higher the temperature is, the hotter it is. Thus the hottest cities will be those whose temperatures are positive.

> When comparing two positive numbers, the greatest is the one that is farthest from 0.

The hottest cities are Ajaccio and Bordeaux (positive temperatures), but 8 is further from 0 than 6, so Bordeaux is the hottest city.

Thus the ranking of cities from the coldest to the hottest in the morning is Nancy, Paris, Ajaccio and Bordeaux.

Exercise : Rank these cities from the coldest to the hottest in the afternoon.

The coldest city is Nancy (the only negative temperature).

The hottest cities are Paris, Ajaccio and Bordeaux (positive temperatures) but 14 (Ajaccio) is further from 0 than 10 (Bordeaux), which in turn is further from 0 than 2 (Paris).

Thus the ranking of cities from the coldest to the hottest in the afternoon is Nancy, Paris, Bordeaux and Ajaccio.

To add two relative numbers:

If they have the same sign, we add their distances to zero and we keep the common sign.

If they are of opposite signs, subtract their distances to zero and take the sign of one which has the greatest distance to zero.

Examples:

If we want to calculate (-5) + (-7) → we add two numbers with the same sign,
Their distances to 0 are added: 5 + 7 = 12,
We keep the common sign which is (-) minus,
Giving -12.

We want to calculate (5) + (-7) → we add two numbers of opposite signs,
Their distances to 0 are subtracted: 7-5 = 2,
We take the sign of the number which has the greatest distance to 0. 7 has a greater distance to 0 than 5, so we keep the sign of -7
Giving -2.

To subtract two relative numbers:

Subtracting a relative number is the same as adding its opposite.

Example:

We want to calculate (-5) - (-7) → we want to subtract the number -7,
We add the opposite of -7 i.e. (-5) + 7
We return to the addition of two numbers of opposite signs → we calculate 7 – 5 = 2 and we keep the sign of the number which has the greatest distance to 0. 7 has a greater distance to 0 than 5, so we keep the sign of 7,
Giving (-5) – (-7) = 2.

Let us return to our problem and start with Ajaccio; the difference between the temperature of the afternoon and morning is given by 14-6 = 8. The temperature therefore increases by 8 °C between the morning and afternoon.

For Bordeaux; the difference between the temperature of the afternoon and morning is calculated by 10-8 = 2. The temperature therefore increases by 2 °C between the morning and afternoon.

For Nancy; the difference between the temperature of the afternoon and morning is the difference between -9 and -1 or (-1) - (-9).

Here we calculate the difference of 2 relative numbers, by applying the rule "Subtracting a relative number is the same as adding its opposite"
$(-1) - (-9) = (-1) + 9$

Now we add 2 relative numbers of opposite signs, by applying the rule « Subtract their distances to zero and take the sign of the one which has the greatest distance to zero. »:

$9 - 1 = 8$ and we keep the sign + (because the distance of 9 to 0 is greater than the distance of -1 to 0)

So $(-1) - (-9) = 8$

We deduce that the temperature increases by 8 °C between the morning and afternoon.

For Paris; the difference between the temperature of the afternoon and morning is the difference between 2 and (-4): 2 - (-4). Here we calculate the difference of 2 relative numbers, by applying the rule "Subtracting a relative number is the same than adding its opposite"
$2 - (-4) = 2 + 4 = 6$.

So the temperature increases by 6 °C between the morning and afternoon.

History and timeline

Many characters have marked history over the centuries. The following table lists the year of birth of some great historical figures:

Name	Year of birth
Napoleon	1769
Ramses II	1303 before JC
Julius Caesar	100 before JC
Louis XIV	1638
Socrates	470 before JC

Exercise: Classify these characters from the oldest to the youngest.

When it comes to dates, the commonly used benchmark is the birth of Jesus Christ (BC) placed in year 0. Therefore, any date before the birth of Jesus Christ will be preceded by a minus sign (-) and any date subsequent to the birth of Jesus Christ will be preceded by a plus sign (+). We can therefore rewrite the table as follows:

Name	Year of birth
Napoleon	+1769
Ramses II	-1303
Julius Caesar	-100
Louis XIV	+1638
Socrates	-470

The older characters are those for which the birth year is negative, i.e. Ramses II (-1303), Julius Caesar (-100) and

Socrates (-470). Among these three characters, the oldest is the one whose birth year has the largest distance to 0: It is Ramses II, followed by Socrates and Julius Caesar.

Among the more recent characters are Napoleon (1769) and Louis XIV (1638), the youngest is the one whose birth year has the largest distance to 0: It is Napoleon.

The classification from the oldest to the youngest figure is as follows:

Name	Year of birth
Ramses II	-1303
Socrates	-470
Julius Caesar	-100
Louis XIV	+1638
Napoleon	+1769

Scuba diving

A team is responsible for preparing a snorkeling competition. Competitors coming up from under the sea must respect decompression stops during their ascent to avoid diving accidents. Therefore 7 buoys will be placed along the route. The place where the competition will take place is -140 meters deep.

Exercise: Determine the depths at which to place the buoys to mark out the course.

The entire route has a depth of -140 meters and must be divided into equal sections by 7 buoys. For the distance between each buoy, perform the following division: -140 / 7.

To calculate the quotient of a relative number to a non-zero relative number, divide their distance to zero and apply the following rule of signs:

The quotient of two relative numbers of the same sign is positive;
The quotient of two relative numbers of opposite signs is negative.

Applying this rule to our example:

Divide the distance to 0 of (-140) and (7): 140/7 = 20,
The two numbers have opposite signs, the result is negative,

So (-140) / 7 = - 20.

This means that a buoy will be placed at -20 m depth, then following at -40 m and -60 m and so on down to - 140 m.

Here are also useful calculation rules for multiplying relative numbers:

To multiply two relative numbers, multiply their distance to zero and apply the following rule of signs:

The product of two relative numbers of the same sign is positive;
The product of two relative numbers of opposite signs is negative.

The product of several relative numbers is:

Positive if it has an even number of negative factors.
Negative if it has an odd number of negative factors.

Examples:

Calculate (-5) x (-4):

We calculate 5 x 4 = 20,
(-5) and (-4) have the same signs, the result is positive,
So (-5) x (-4) = 20.

Calculate (-6) x (3):

We calculate 6 x 3 = 18,
(-6) and 3 have opposite signs, the result is negative,
So (-6) x 3 = - 18.

Calculate (-6) x (-5) x (4):

We calculate 6 x 5 x 4 = 120,
The multiplication has a positive number (4) and **two** negative numbers (-6 and -5) → even number of negative factors: the result is positive,
So (-6) x (-5) x (4) = 120.

Calculate (-3) x (-5) x (-2):

We calculate 3 x 5 x 2 = 30,
The multiplication has **three** negative numbers (-3, -5 and -2) → **odd number** of negative factors: the result is negative,
So (-3) x (-5) x (-2) = -30.

Squares and square roots

What is the square of a number?

Road safety

Braking distance is the distance required to stop a vehicle with the brakes. It depends on the speed and road conditions (dry or wet).

We can calculate the distance using the formula:

$$D = k \times V^2$$

With:
D is the distance required for braking (in meters),
k is a coefficient which is 0.0048 for a dry road and 0.0098 for a wet road,
V is the vehicle speed (in km / h).

Exercise: Calculate the braking distance of a car at 90 km/h on a dry road and then on a wet road.

In the formula above, you may notice a little ^{2} above V. Thus V^2 reads: « V squared ».

The **square** of a number A is written A^2.
To calculate A^2 , multiply A x A

For example: 5^2 = 5 x 5 = 25 and 4^2 = 4 x 4 = 16 and 7^2 = 7 x 7 = 49.

The first calculation we have to make is for a car on dry roads:

We will use the 0.0048 value for the coefficient k: k = 0.0048.

The speed of the car is 90 km/h so V = 90.

With the formula above: D = 0.0048 x 90^2 or D = 0.0048 x 90 x 90.

It gives D = 38.88; D is in meters so D = 38.88 meters.

This means that a car traveling at 90 km/h on a dry road will need about 40 meters to stop from the moment when the driver presses the brake pedal.

The second calculation concerns a car traveling at 90 km/h (V = 90) on a wet road, this is why:

We will use now the 0.0098 value for the coefficient k: k = 0.0098.

The above formula gives us: D = 0.0098 x 90^2, written also as; D = 0.0098 x 90 x 90.

Which gives D = 79.38 meters.

This means that a car traveling at 90 km/h on a wet road will need about 80 meters to stop from the moment when the driver presses the brake pedal.

It can be seen that a car traveling at 90 km/h will need a longer braking distance on wet roads than on dry roads to stop. This explains why the speed limits may be different on the road depending on weather conditions.

What is the square root of a number?

Road safety (continued)

Exercise: A driver leaves a braking distance of 20 meters in front of him. Calculate the maximum speed at which he can drive on dry and wet road to avoid the risk of accidents.

Let us consider first the case where the driver is on a dry road in these conditions:

k = 0.0048
D = 20 meters (the maximum braking distance without risking hitting the car in front of it).

By replacing these components in the formula, we get:
$20 = 0.0048 \times V^2$.

We are looking for V^2:

Divide the right expression ($0.0048 \times V^2$) by 0.0048 for having only V^2;
In this case, we must also divide the left term by 0.0048 to keep the calculation consistent on both sides.

So we can write that $V^2 = 20 / 0.0048$.

So $V^2 = 4167$.

So $V = \sqrt{4167}$

For example:
√49 = 7 because 7 x 7 =49; or √25 = 5 because 5 x 5 = 25; or √64 = 8 because 8 x 8 = 64.

To calculate a square root, there is a specific key on the calculator, simply enter the number (here 4167) and then press the √ key.
It is found that V = 64.5 km/h.

Thus a driver leaving only 20 meters in front of him can drive at 64 km/h maximum on a dry road without collision risk.

Make the same calculation if the driver is traveling now on a wet road, in these conditions:

k = 0.0098 (coefficient for wet road)
D = 20 meters (the maximum braking distance without risking hitting the car in front of it).

By replacing these components in the formula, we get:
$20 = 0.0098 \times V^2$.

So $V^2 = 20 / 0.0098$.

So $V^2 = 2040.8$.

And V = $\sqrt{2040.8}$

So V = 45 km/h (round).

Thus a driver leaving only 20 meters in front of him can drive at 45 km/h maximum on a wet road without collision risk.

Power of a number

Decimal numbers are very common in everyday life because they are fully suitable for various uses. So to fix the price of an object, we need at most two decimal places. If I speak about the price of bread, a price of € 0.95 will be familiar but you will consider me as an eccentric guy if I tell you that a loaf of bread costs € 0.9562145 ... When we measure the size of an object or a person, we rarely need more than two decimal places. You might also be astonished if the doctor tells you that your child measures 1.4345874 meters, whereas if he announces a height of 1.43 meters, you will find it quite normal. However, there are subjects for which we have to use very large or conversely very small numbers. In these situations, the numbers can be extremely long. That's why we created powers of numbers.

What is a power?

We must first define what a power is:

We saw in the previous chapter that 5^2 (five square) corresponds to 5 x 5. 5^2. We can also say "5 to the Power of 2".

Similarly, 5^3 (five cubed) corresponds to 5 x 5 x 5 and 5^3 may also say "5 to the power of 3".

Similarly, 5^6 correspond to 5 x 5 x 5 x 5 x 5 x 5 and 5^6 called "5 to the power of 6".

A^n is called « **A power n** » and is a multiplication with « n times » number A.

Ex. 4^5 (4 to the power of 5) is to multiply 5 times number 4 (4 x 4 x 4 x 4 x 4)

Vocabulary:
A^2 says « A squared »
A^3 says « A cubed »

Powers of 10

One of the most used powers is the power of 10. Let's look at what powers of 10 look like:

$10^2 = 10 \times 10 = 100$
$10^3 = 10 \times 10 \times 10 = 1000$
$10^4 = 10 \times 10 \times 10 \times 10 = 10000$
And so on...

We notice that 10^n is the number 1 followed by n times the number 0.

10^n is the number 1 followed by n times the number 0 (10 power n).

Astronomy

One of the favorite areas of power usage is astronomy. Indeed, when we talk about planets, the distances become immediately gigantic to the point where it is painful to quickly transcribe.

Consider for example the distance from Earth to the Sun; it is 149.6 million kilometers. This distance can be written as 149600000 km, which makes a lot of 0s ... To simplify this, scientists use powers of 10.

In our example, we have the number 149600000.

This number can be written as 1496 x 100000. Why? Answer in the next chapter.

How to multiply by a power of 10?

Astronomy (continued)

149600000 can be written as 1496 x 100000 because:

To multiply a number by 10, just add a 0,
To multiply a number by 100, just add two 0
To multiply a number by 1000, just add three 0
To multiply a number by 10,000, just add four 0
To multiply a number by 10000000000, just add ten 0.

To multiply a number by 10^n, just add n 0.

Thus in the number 149600000, 1496 was followed by five 0, this corresponds to 1496 multiplied by 100,000.

Furthermore 100000 is the number 1 followed by five zeros, it can be written 10^5.

Thus 149600000 = 1496 x 100 000 = 1496 x 10^5

We wrote 149600000 as a number (1496) multiplied by a power of 10 (10^5), which simplifies the writing.

But scientists are not short on innovations and have created a universal notation to write very large numbers according to a defined standard, it is the "scientific notation."

What is scientific notation?

Scientific notation is to write a number as a decimal number with one digit from 1 to 9, before the decimal point and multiplied by a power of 10.

Ex.
4,874 x 10^3 is a scientific notation,
54.87 x 10^4 is not a scientific notation because 54.87 is not a number with one digit before the decimal point,
3.21 x 100 is not a scientific notation because 100 is not expressed as a power of 10 (10^2).

Astronomy (continued)

We have previously written that 149600000 = 1496 x 10^5, but how do we write this in the form of a scientific notation?

Indeed, this number is not yet a scientific notation as the number that appears before the power of 10 (1496) is not a decimal number with a single digit between 1 and 9, before the decimal point.

1.496 is, meanwhile, a decimal number with a single digit between 1 and 9, before the decimal point.

So, how to go from 1496 to 1.496?

We see that starting from 1.496 if we can shift the decimal point three places to the right, we get 1496.

To move the decimal point one place to the right, multiply the number by 10 or 10^1;
To move the decimal point two places to the right, multiply the number by 100 or 10^2;
To move the decimal point three places to the right, multiply the number by 1000 or 10^3;

And $1496 = 1.496 \times 1000 = 1.496 \times 10^3$

To shift the decimal point of a decimal number n places to the right, you have to multiply by 10^n

Multiplying a number by 10^n amounts to shifting the point of n places to the right.

In summary, we have seen that $149600000 = 1496 \times 100000 = 1496 \times 10^5$ and $1496 = 1,496 \times 1000 = 1,496 \times 10^3$.

So we can write $149600000 = 1,496 \times 10^3 \times 10^5$.

It is similar to scientific notation, however, if it is necessary to perform the multiplication of two powers of 10 ($10^3 \times 10^5$): then how can we multiply these two powers of 10?

How to multiply two powers of 10?

Here is the rule for multiplication of two powers of 10:

To multiply two powers of the same number A:

$$A^n \times A^m = A^{n+m}$$

Ex. $10^3 \times 10^5 = 10^{3+5} = 10^8$

Astronomy (continued)

We can write:

$$149600000 \quad = 1{,}496 \times 10^3 \times 10^5$$
$$= 1{,}496 \times 10^{(3+5)}$$
$$= 1{,}496 \times 10^8$$

The distance from the Earth to the Sun is $1{,}496 \times 10^8$ km, and this distance is well expressed in the form of scientific notation.

Here are the distances from the Sun to different planets of the solar system:

93

Planet	Distance to Sun
Mercury	58 million km
Jupiter	0.775 billion km
Venus	108 million km
Pluto	49 billion km
Mars	228 million km
Neptune	4.504 billion km
Saturn	1429 million km
Uranus	2.869 billion km

Exercise: Express these distances in scientific notation and then classify them from farthest to closest to the sun.

To solve this exercise, we need information about large numbers, as follows:

1 million is 1000000 or 10^6 (there are six 0),
1 billion is 1000000000 or 10^9 (there are nine 0).

Mercury is located at 58 million kilometers so: 58×10^6 km.
It is not a scientific notation as 58 is not a number between 1 and 9. We have to change 58 into a number between 1 and 9.
Note that $58 = 5.8 \times 10$ and $10 = 10^1$.

We can write $58 \times 10^6 = 5.8 \times 10^1 \times 10^6$
Considering the rule above: $10^1 \times 10^6 = 10^{1+6} = 10^7$
So $58 \times 10^6 = 5.8 \times 10^{7,}$
Here we have a scientific notation to Mercury's distance to the sun.

Proceed the same way for Jupiter which is located at 0.775 billion km from the Sun.

We can write: $0{,}775 \times 10^9$ km.

It is not a scientific notation as 0.775 is not a number between 1 and 9.

We have to change 0.775 into a number between 1 and 9.

Note that $0.775 = 7.75 / 10$.

How to divide by a power of 10?

Astronomy (continued)

Here is a very useful property:

To divide a number by 10, you have to shift the decimal point one place to the left,
To divide a number by 100, shift the decimal point two places to the left,
To divide a number by 1000, shift the decimal point three places to the left
To divide a number by 10,000, shift the decimal point four places to the left,
To divide a number by 10000000000, shift the decimal point ten places to the left.

To divide a number by 10^n, shift the decimal point n places to the left.

Let us pause in the resolution of astronomical exercises to illustrate the concept above:

Consider the number 125.48:

Dividing this number by 10 shifts the decimal point one place to the left, so 125.48 / 10 = 12.548.

Dividing this number by 100 amounts to shifting the decimal point two places to the left, so 125.48 / 100 = 1.2548.

Dividing this number by 1000000 amounts to shifting the decimal point six places to the left. Here it seems impossible since once we shifted the decimal point three places, there are no more numbers ... To do this, just simply add 0. Thus 125.48 / 100000 = 0.00012548.

Note that:

125.48 / 10 can be written 125.48 / 10^1
125.48 / 100 can be written 125.48 / 10^2
125.48 / 1000000 can be written 125.48 / 10^6

Dividing a number by 10^n is equivalent to multiplying that number by 10^{-n}.

So,

125.48 / 10 = 125.48 / 10^1 = 125.48 x 10^{-1}
125.48 / 100 = 125.48 / 10^2 = 125.48 x 10^{-2}
125.48 / 1000000 = 125.48 / 10^6 = 125.48 x 10^{-6}

Remember the following principles:

For multiplying a number by a positive power of 10 (10^n), the point of the number is shifted n places to the right.

For multiplying a number by a negative power of 10 (10^{-n}), the point of the number is shifted n places to the left.

This property is deduced that:

$125.48 \times 10^{-1} = 12.548$ (we shift the decimal point one place to the left)
$125.48 \times 10^{-2} = 1.2548$ (we shift the decimal point two places to the left)
$125.48 \times 10^{-6} = 0.00012548$ (we shift the decimal point 6 places to the left).

By using the last 3 rules described above:
$125.48 / 10^1 = 125.48 \times 10^{-1} = 12.548$
$125.48 / 10^2 = 125.48 \times 10^{-2} = 1.2548$
$125.48 / 10^6 = 125.48 \times 10^{-6} = 0.00012548$

Consider the number 48 and calculate 48 / 10, then 48 / 100, then 48 / 10000.

With the above properties, we can write:

$48 / 10 = 48 / 10^1 = 48 \times 10^{-1}$
$48 / 100 = 48 / 10^2 = 48 \times 10^{-2}$
$48 / 100000 = 48 / 10^5 = 48 \times 10^{-5}$

However, in the number 48, there is no decimal point! In this case, consider that the decimal point is located right after 8. Indeed, 48 is the same as 48.0 or 48.00 or 48.0000.

So;

$48/10 = 48/10^1 = 48 \times 10^{-1} = 4.8$ (we shift the decimal point one place to the left)
$48/100 = 48/10^2 = 48 \times 10^{-2} = 0.48$ (we shift the decimal point two places to the left)
$48/100000 = 48/10^5 = 48 \times 10^{-5} = 0.00048$ (we shift the decimal point 5 places to the left)

One last property on the powers:

A number to the power of 0 (A^0) is always equal to 1.

Ex : $10^0 = 4^0 = 13^0 = 1541^0 = 1$

We are now equipped to return to our exercise on the planets in which we had to write the distance to the Sun of each of the planets in scientific notation.

For Jupiter, we had to write 0.775×10^9 km with scientific notation and had noticed that $0.775 = 7.75 / 10$.

With the last properties addressed, we can write that $0.775 = 7.75 / 10 = 7.75 / 10^1 = 7.75 \times 10^{-1}$

We can write $0.775 \times 10^9 = 7.75 \times 10^{-1} \times 10^9$ or $7.75 \times 10^{(-1+9)} = 7.75 \times 10^8$ and it is a scientific notation.

Can you follow the same reasoning for Venus, Pluto, Mars, Neptune, Saturn and Uranus?

Venus is located at 108 millions km from the Sun: 108×10^6 km.
$108 = 1.08 \times 100 = 1.08 \times 10^2$
So $108 \times 10^6 = 1.08 \times 10^2 \times 10^6 = 1.08 \times 10^{(2+6)} = 1.08 \times 10^8$

Pluto is located at 49 billion km from the Sun: 49×10^9 km.
$49 = 4.9 \times 10 = 4.9 \times 10^1$
So $49 \times 10^9 = 4.9 \times 10^1 \times 10^9 = 4.9 \times 10^{(1+9)} = 4.9 \times 10^{10}$

Mars is located at 228 millions km from the Sun: 228×10^6 km.
$228 = 2.28 \times 100 = 2.28 \times 10^2$
Donc $228 \times 10^6 = 2.28 \times 10^2 \times 10^6 = 2.28 \times 10^{(2+6)} = 2.28 \times 10^8$

Neptune is located at 4,504 billion km from the Sun: 4.504×10^9 km.
It is a scientific notation.

Saturn is located at 1429 millions km from the Sun: 1429×10^6 km.
$1429 = 1.429 \times 1000 = 1.429 \times 10^3$
So $1429 \times 10^6 = 1.429 \times 10^3 \times 10^6 = 1.429 \times 10^{(3+6)} = 1.429 \times 10^9$

Uranus is located at 2.869 billion km from the Sun: 2.869 x 10^9 km.
It is a scientific notation.

All distances of the planets from the Sun are now expressed in scientific notation, which gives:

Planet	Distance from the Sun (km)
Mercury	5,8 x 10^7
Jupiter	7,75 x 10^8
Venus	1,08 x 10^8
Pluto	4,9 x 10^{10}
Mars	2,28 x 10^8
Neptune	4,504 x 10^9
Saturn	1, 429 x 10^9
Uranus	2,869 x 10^9

We must now rank these planets from the farthest to the closest to the Sun.

How to compare powers of 10?

Astronomy (continued)

> **To compare two numbers in scientific notation:**
>
> **The largest is the one which has the highest power of 10.**
>
> **If the powers of 10 are identical, the largest is the one with the highest number before the power.**
>
> *Ex.*
> *2.1×10^7 is greater than 8.7×10^6 because 10^7 is greater than 10^6*
>
> *5.4×10^5 is greater than 4.004×10^5 as the powers of 10 are identical (10^5), and 5.4 is greater than 4.004.*

The largest power of 10 is for Pluto (10^{10}). So this is the farthest planet from the sun.

Neptune, Saturn and Uranus both have a distance of about 109. We have to compare the figures before the power: 4.504, 1.429 and 2.869. We deduce that Neptune is farther from the Sun than Uranus and Saturn.

Jupiter, Venus and Mars all have a distance of about 10^8. We compare the figures before the power: 7.75, 1.08

and 2.28. We deduce that Jupiter is farther from the Sun than Mars and Venus.

This gives the ranking of the planets from the farthest to the closest to the Sun:

Planet	Distance from the Sun (km)
Pluto	$4,9 \times 10^{10}$
Neptune	$4,504 \times 10^{9}$
Uranus	$2,869 \times 10^{9}$
Saturn	$1,429 \times 10^{9}$
Jupiter	$7,75 \times 10^{8}$
Mars	$2,28 \times 10^{8}$
Venus	$1,08 \times 10^{8}$
Mercury	$5,8 \times 10^{7}$

We have seen that the distance from the Earth to the Sun is $1,496 \times 10^{8}$ km which gives the following final table:

Planet	Distance from the Sun (km)
Pluto	$4,9 \times 10^{10}$
Neptune	$4,504 \times 10^{9}$
Uranus	$2,869 \times 10^{9}$
Saturn	$1,429 \times 10^{9}$
Jupiter	$7,75 \times 10^{8}$
Mars	$2,28 \times 10^{8}$
Earth	$\mathbf{1,496 \times 10^{8}}$
Venus	$1,08 \times 10^{8}$
Mercury	$5,8 \times 10^{7}$

Your turn!

These 100 pages promised to make peace with decimal numbers, relative numbers, squares and square roots, fractions and powers and to show how they are reached. I propose now to practice on all the concepts that were discussed throughout the book.

In the following pages, small exercises help you to review and check your understanding of the different chapters of the book. Take time to solve the exercises before you read the corrections. Please read the relevant chapter and you will realize that you are now much more able to tackle mathematics than you were some pages before.

Decimal numbers

Exercise 1

Sort the following numbers in descending order:
0.458; 1.5874; 1.5847; 0.4589; 0.98746; 0.97; 0.4

Exercise 2

Set the addition and calculate:
a/ 12.587 + 2.9845
b/ 0.548 + 2.413658

Exercise 3

Perform the following conversions:
a/ 256 g in kg
b/ 23 mg in g
c/ 0.1 kg in mg
d/ 12 L in cL
e/ 61.7 dL in mL

Exercise 4

Determine, without performing the calculation, the number of decimal places in the result of the following multiplications:
a/ 12.547 x 1.6587
b/ 6.23 x 3.1
c/ 0.123 x 1.6
d/ 87458.45 x 9854.11

Exercise 5

Set the multiplications and calculate:

a/ 14.51 x 11.2
b/ 6.253 x 4.1

Exercise 6

Determine by what number(s) between 2 and 13, the following numbers are divisible:
a/ 462
b/ 240
c/ 891

Exercise 7

I want to divide 1485 by 7:
a / what number is the dividend?
b / what number is the divisor?
c / what is the quotient?
d / what is the remainder?

Same questions: If I want to divide 1255 by 5?

Relative numbers

Exercise 8

Sort the following numbers in ascending order:
125 ; 158 ; -8.5 ; -12 ; -12.3 ; 50.1 ; -0.9 ; 0.4

Exercise 9

Perform the following calculations:
a/ -24 - 13
b/ 21 – 48
c/ -12 + 23
d/ -5 + 8

Exercise 10

What will be the sign in the result of each following calculation?
a/ -12 x 26 x (-2) x (-14)
b/ 21 x 7 x (-1) x (-2)
c/ -12 x 23 x 6 x 5 x (-4) x (-9) x 2 x (-22)
d/ 15 x 11 x 9 x 6 x 2 x (-4)

Exercise 11

Perform the following multiplications:
a/ (-2) x (-6) x 4
b/ 2 x (-3) x (-5)
c/ 11 x 7 x (-2)

Fractions

Exercise 12

Write the following numbers in the form of fractions:
a / six ninths
b / four twelfths
c / twenty five hundredths
d / two fifteenths
e / one hundred and ten two hundred twentieths

Exercise 13

Express :
a / 2/5 in a fraction whose denominator is 15
b / 3/4 in a fraction whose denominator is 24
c / 6/11 in a fraction whose denominator is 77
d / 7/10 in a fraction whose denominator is 100

Exercise 14
Compare the following fractions:
a / 7/8 and 2/3
b / 12/5 and 41/15
c / 6/7 and 18/21
d / 5/9 and 7/11

Exercise 15

Perform the following additions:
a/ 4/5 + 1/3 + 7/30
b/ 2/11 + 3/2 + 13/3

Exercise 16

Perform the following multiplications:
a/ 3/5 x 6/5
b/ 7/11 x 9/4
c/ 3/14 x 5/2

Exercise 17

Simplify, if possible, the following fractions:
a/ 12/36
b/ 56/88
c/ 9/14

Exercise 18

I have a pie:
c / If I eat 5/9, what is the remainder?
b / If I eat 11/14, what is the remainder?
c / If I eat 7/16, what is the remainder?

Exercise 19

What is:
a/ 3/4 of 5/8 ?
b/ 1/6 of 7/10 ?
c/ 4/5 of 3/4 ?

Squares and square roots

Exercise 20

Determine the following squares:

a/ 6^2

b/ 8^2

c/ 11^2

d/ 14^2

Exercise 21

Determine the following square roots:

a/ 25

b/ 1764

c/ 441

d/ 289

Powers

Exercise 22

Write the following numbers in the form of a power of 10:

a/ 100
b/ 10
c/ 0
d/ 1
e/ 10000
f/ 10000000
g/ 0.1
h/ 0.0001
i/ 0.00000001

Exercise 23

Perform the following multiplications:

a/ $10^3 \times 10^4$
b/ $10^{-9} \times 10^{-5}$
c/ $10^{-4} \times 10^6 \times 10^{-1}$
d/ $10^{-6} \times 10^{11} \times 10^2$

Exercise 24

Perform the following divisions:

a/ $10^5 / 10^2$
b/ $10^{-2} / 10^{-5}$
c/ $10^{-4} / 10^6$
d/ $10^{11} \times 10^{-6}$

Exercise 25

Write the following numbers in scientific notation:
a/ 0.54875
b/ 0.00054699
c/ 15488745
d/ 14589.547

Solutions

Exercise 1

Rank these numbers in descending order to rank them from the largest to the smallest.
The given numbers are decimal numbers (decimal point) with not the same number of digits.

One trick to rank them is to place them in a table always ensuring that the decimal point is placed in the same column as shown below:

0.	4	5	8		
1.	5	8	7	4	
1.	5	8	4	7	
0.	4	5	8	9	
0.	9	8	7	4	6
0.	9	7			
0.	4				

All blanks can be supplemented with 0, which gives:

0.	4	5	8	0	0
1.	5	8	7	4	0
1.	5	8	4	7	0
0.	4	5	8	9	0
0.	9	8	7	4	6
0.	9	7	0	0	0
0.	4	0	0	0	0

To compare numbers, you have just to look at the table from left to right.

In the 1st column, the highest digit is 1. We deduce the highest number is either 1.5874 or 1.5847. To decide, look at the 2nd column, for each number there is a 5 in the 2nd column, so it is not possible to conclude yet. Look at the 3rd column, for each number there is an 8 in the 3rd column; so it is still impossible to conclude. Look at the 4th column, for the compared numbers, there are a 7 and a 4. 7 is larger than 4, so 1.5874 is larger than 1.5847.

For the 5 others numbers to rank, there is a 0 in the 1st column. So we have to look the 2nd column to compare them. In this 2nd column, the highest digit is 9 for numbers 0.98746 and 0.97. To compare these two numbers, look at the 3rd column. In this column, there are an 8 and a 7. We deduce that 0.98746 is larger than 0.97.

Proceeding in the same way for the remaining numbers, we deduce the following classification:

1.5874; 1.5847; 0.98746; 0.97; 0.4589; 0.458; 0.4

Exercise 2

Performing operations on decimal numbers is difficult because of the presence of a decimal point.

One trick to rank them is to place them in a table always ensuring that the decimal point is placed in the same column as shown below:

1	2.	5	8	7	
	2.	9	8	4	5

Then fill in all the blanks with 0, which gives:

1	2.	5	8	7	0
0	2.	9	8	4	5

Now perform the calculation as a classic addition, column by column, starting with the right and paying attention to numbers to carry over:

1	2.	5	8	7	0
0	2.	9	8	4	5
1	5.	5	7	1	5

We deduce that 12.587 + 2.9845 = 15.5715.

Similarly, we can find that 0.548 + 2.413658 = 2.961658.

Exercise 3

To perform conversions, it is strongly recommended to learn and know how to reproduce the following tables:

kg	hg	dag	g	dg	cg	mg
kilogram	hectogram	decagram	gram	decigram	centigram	milligram

kL	hL	daL	L	dL	cL	mL
kiloliter	hectoliter	decaliter	liter	deciliter	centiliter	milliliter

Then, simply place the given numbers to convert in this table ensuring that the digit followed by the decimal point is in the box corresponding to the unit to convert. If the number does not contain a decimal point, it will be considered that it is after the last digit of the number.

For example to convert 256 g, it is considered that the decimal point is after the 6 (although it does not appear), resulting in the table:

kg	hg	dag	g	dg	cg	mg
	2	5	6.			

We wish to convert it in kilograms, which is possible by adding 0 as necessary to reach the kg box. Furthermore, we shift the decimal point to the digit located in the kg box, which gives:

kg	hg	dag	g	dg	cg	mg
0.	2	5	6			

That means 256 g = 0.256 kg.

Similarly we can convert 23 mg into g:

kg	hg	dag	g	dg	cg	mg
					2	3.

There is no decimal point in the number 23; therefore it is considered that the decimal point is after the digit 3 and we write the number in the table so that the digit 3 is in the mg box. Then 0 are added to reach the g box and the decimal point is moved there, which gives:

kg	hg	dag	g	dg	cg	mg
			0.	0	2	3

So 23 mg = 0.023 g.

The procedure is the same for the following conversions:

Conversion of 0.1kg in mg:

kg	hg	dag	g	dg	cg	mg
0	1	0	0	0	0	0.

So 0.1 kg = 100000 mg.

Conversion of 12 L into cL:

kL	hL	daL	L	dL	cL	mL
		1	2	0	0.	

So 12 L = 1200 cL

Conversion of 61.7 dL into mL:

kL	hL	daL	L	dL	cL	mL
			6	1	7	0.

So 61.7 dL = 6170 mL

Exercise 4

To determine, without calculating, the maximum number of decimal places in the result of multiplication, just count the number of decimal places in both numbers to multiply and then add up the two numbers thus obtained. For example:

In operation 12.547 x 1.6587, the number 12.547 has 3 decimal places and 1.6587 has 4 decimal places.

The result of this multiplication will have 3 + 4 = 7 decimal places.

12.547 x 1.6587 = 20.8117089

Similarly, the result of 6.23 x 3.1 will have 2 + 1 = 3 decimal places.

6.23 x 3.1 = 19.313

Similarly, the result of 0.123 x 1.6 will have 3 + 1 = 4 decimal places.

0.123 x 1.6 = 0.1968

To finish, the result of 87458.45 x 9854.11 will have 2 + 2 = 4 decimal places.

87458.45 x 9854.11 = 861825186.7295.

Exercise 5

To perform a multiplication as 14.51 x 11.2, you should write the calculation with the same number of decimal places (complete with 0), ie: 14.51 x 11.20.

Then, set and multiply regardless of the decimal point, as follows:

```
            1   4   5   1
        x   1   1   2   0
            0   0   0   0
        2   9   0   2   .
    1   4   5   1   .   .
1   4   5   1   .   .   .
1   6   2   5   1   2   0
```

As seen in the previous exercise, the result of 14.51 x 11.20 will have 2 + 2 = 4 decimal places.

Starting from the last digit of the result, count 4 digits and place the decimal point, giving 162.5120.

We deduce that 14.51 x 11.2 = 162.512.

Similarly, 6.253 x 4.1 x 4.100 = 6.253.
We set the multiplication regardless of decimal points and we know that the result will have 3 + 3 = 6 decimal places.

We find 4.1 x 6.253 = 25.637300; you can delete the 0 in the end giving 6.253 x 4.1 = 25.6373 (ie 3 + 1 = 4 decimal places).

119

Exercise 6

The following table lists the criteria for divisibility presented in the book applied to each of the numbers proposed in the exercise:

Divisibility criteria	Example
Divisibility by 2: A number is divisible by 2 if its last digit is even (it finishes with 0, 2, 4, 6 or 8)	462 is divisible by 2 240 is divisible by 2 891 is not divisible by 2
Divisibility by 3: A number is divisible by 3 if the addition of its digits is a multiple of 3.	462 is divisible by 3 as 4+6+2=12 and 12 is a multiple of 3 240 is divisible by 3 as 2+4+0=6 and 6 is a multiple of 3 891 is divisible by 3 as 8+9+1=18 and 18 is a multiple of 3
Divisibility by 4: A number is divisible by 4 if its two last digits are a multiple of 4.	462 is not divisible by 4 as 62 is not a multiple of 4 240 is divisible by 4 as 40 is a multiple of 4 891 is not divisible by 4 as 91 is not a multiple of 4

Divisibility by 5: A number is divisible by 5 if its last digit is a 0 or a 5.	462 is not divisible by 5 as the last digit is 2 240 is divisible by 5 as the last digit is 0 891 is not divisible by 5 as last digit is 1
Divisibility by 6: A number is divisible by 6 if it is at the same time divisible by 2 and by 3.	462 is divisible by 6 as it is divisible by 2 and 3 240 is divisible by 6 as it is divisible by 2 and 3 891 is not divisible by 6 as it is not divisible by 2 and 3
Divisibility by 7: A number is divisible by 7 if ht − (ux2) is divisible by 7. In the number: h is the digit of hundreds, t the digit of tens and u the digit of units.	462 is divisible by 7 as ht − (u x 2) = 46 − 2x2 = 46-4 = 42 is divisible by 7 240 is not divisible by 7 as ht − (u x 2) = 24 − 0x2 = 24-0 = 24 is not divisible by 7 891 is not divisible by 7 as ht − (u x 2) = 89 − 1x2 = 89-2 = 87 is not divisible by 7

121

Divisibility by 8: A number is divisible by 8 if ht + (u/2) is divisible by 4. In the number: h is the digit of hundreds, t the digit of tens and u the digit of units.	462 is not divisible by 8 as ht + (u/2) = 46 + (2/2) = 47 is nor divisible by 4 240 is divisible by 8 as ht + (u/2) = 24 + (0/2) = 24 is divisible by 4 891 is not divisible by 8 as ht + (u/2) = 89 + (1/2) = 89,5 is not divisible by 4
Divisibility by 9: A number is divisible by 9 if its digit addition is a multiple of 9.	462 is not divisible by 9 as 4+6+2=12 and 12 is not divisible by 9 240 is not divisible by 9 as 2+4+0=6 and 6 is not divisible by 9 891 is divisible by 9 as 8+9+1=18 and 18 is divisible by 9
Divisibility by 10: A number is divisible by 10 if its last digit is 0.	462 is not divisible by 10 as the last digit is 2 240 is divisible by 10 as the last digit is 0. 891 is not divisible by 10 as the last digit is 1

122

Divisibility by 11: A number is divisible by 11 if the difference between the even digit addition and the odd digit addition is divisible by 11.	4**62** is divisible by 11 as 4+2=6 and 6-6=0 is divisible by 11 **24**0 is not divisible by 11 as 2+0=2 and 4-2=2 is not divisible by 11 8**91** is divisible by 11 as 8+1=9 and 9-9=0 is divisible by 11
Divisibility by 12: A number is divisible by 12 if it is at the same time divisible by 3 and by 4.	462 is not divisible by 12 as it is not divisible by 4 (but it is divisible by 3) 240 is divisible by 12 as it is divisible by 3 and 4 891 is not divisible by 12 as it is not divisible by 4 (but it is divisible by 3)
Divisibility by 13: A number is divisible by 13 if ht + (4xu) is divisible by 13. In the number: h is the digit of hundreds, t the digit of tens and u the digit of units.	462 is not divisible by 13 as 46 + 4x2 = 46+8=54 is not divisible by 13 240 is not divisible by 13 as 24 + 4x0 = 24+0=24 is not divisible by 13 891 is not divisible by 13 as 89 + 4x1 = 89+4 = 93 is not divisible by 13

Exercise 7

I want to divide 1485 by 7:

123

a/ the dividend is 1485
b/ the divisor is 7
c/ the quotient is 212
d/ the remainder is 1 (1485 = 212 x 7 + 1)

Similarly for 1255 divided by 5:

a/ the dividend is 1255
b/ the divisor is 5
c/ the quotient is 251
d/ the remainder is 0 (1255 = 251 x 5 + 0)

Exercise 8

To rank these numbers in ascending order you have to write them from the smallest to the largest:

125 ; 158 ; -8.5 ; -12 ; -12.3 ; 50.1 ; -0.9 ; 0.4

Consider each number as a temperature. The coldest temperature is negative and has the greatest value (ex. -50°C is colder than -25°C).

The largest negative number is -12.3 then -12 then -8.5 and finally -0.9.

Then we have the positive numbers. The closest number to 0 is 0.4 then 50.1 then 125 and finally 158.

This leads to the following classification:

-12.3 ; -12 ; -8.5 ; -0.9 ; 0.4 ; 50.1 ; 125 ; 158

Exercise 9

To perform calculations on relative numbers, the following rules are used:

To add two relative numbers:
If they have the same sign, we add their distances to zero and we keep the common sign.
If they are of opposite signs, subtract their distances to zero and take the sign of one which has the greatest distance to zero.

To subtract two relative numbers:
Subtracting the relative number is the same as adding its opposite.

Calculate $-24 - 13$ is the same as $(-24) + (-13) \rightarrow$ we add relative numbers with the same sign,
We add their distance to 0: $24 + 13 = 37$,
We keep the common sign: minus (-),
It gives -37.

Calculate $21 - 48$ is the same as $21 + (-48) \rightarrow$ we add relative numbers of opposite signs,
We subtract their distance to 0: $48 - 21 = 27$,
We take the sign of the one that has the greatest distance to 0: minus (-) from (-48),
It gives -27.

Calculate $-12 + 23$ is the same as adding relative numbers of opposite signs,
We subtract their distance to 0: $23 - 12 = 11$,
We take the sign of the one that has the greatest distance to 0: (+) from (23),
It gives 11.

Calculate -5 + 8 is the same as adding relative numbers of opposite signs.

We subtract their distance to 0: 8 - 5 = 3,

We take the sign of the one that has the greatest distance to 0: (+) de (8),

It gives 3.

Exercise 10

To determine the sign of the result of a multiplication with relative numbers, we use the following rule:

The result of a multiplication with relative numbers is:

Positive if the multiplication has a even number of negative factors.
Negative if the multiplication has an odd number of negative factors.

In -12 x 26 x (-2) x (-14), there are 3 negative factors (-12 ; -2 ; -14) this is an odd number of negative factors, the result of the multiplication is negative.

In 21 x 7 x (-1) x (-2), there are 2 negative factors (-1 ; -2) this is an even number of negative factors, the result of the multiplication is positive.

In -12 x 23 x 6 x 5 x (-4) x (-9) x 2 x (-22), there are 4 negative factors (-12; -4; -9; -22) this is an even number of negative factors, the result of the multiplication is positive.

In 15 x 11 x 9 x 6 x 2 x (-4), there is 1 negative factor (-4) this is an odd number of negative factors, the result of the multiplication is negative.

Exercise 11

To perform the multiplication (-2) x (-6) x 4, we start by calculating 2 x 6 x 4 without looking at the signs: 2 x 6 x 4 = 12 x 4 = 48.
In this multiplication, there are 2 negative factors (-2 ; -6) this is an even number of negative factors, the result of the multiplication is positive.

So (-2) x (-6) x 4 = 48.

Similarly to calculate 2 x (-3) x (-5), we start by calculating 2 x 3 x 5 without looking at the signs: 2 x 3 x 5 = 6 x 5 = 30.
In this multiplication, there are 2 negative factors (-3 ; -5) this is an even number of negative factors, the result of the multiplication is positive.

So 2 x (-3) x (-5) = 30.

To calculate 11 x 7 x (-2), we start by calculating 11 x 7 x 2 without looking at the signs: 11 x 7 x 2 = 77 x 2 = 154.
In this multiplication, there is 1 negative factor (-2) this is an odd number of negative factors, the result of the multiplication is negative.

So 11 x 7 x (-2) = -154.

Exercise 12

a/ six ninths = 6/9

127

b/ four twelfths = 4/12
c/ twenty five hundredths = 25/100
d/ two fifteenths = 2/15
e/ one hundred and ten two hundred twentieths = 110/220

Exercise 13

a/ To express 2 / 5 in a fraction whose denominator is 15, multiply the denominator (5) by 3 as 5 x 3 = 15.
But to not change a fraction, if we multiply the denominator by 3, you also multiply the numerator by 3 which gives:

2 / 5 = (2 x 3) / (5 x 3) = 6 / 15

b/ Similarly, to express 3 / 4 in a fraction whose denominator is 24, multiply the denominator (4) by 6 as 4 x 6 = 24.
To not change the fraction, if we multiply the denominator by 6, you also multiply the numerator by 6 which gives :

3 / 4 = (3 x 6) / (4 x 6) = 18 / 24

c/ To express 6 / 11 in a fraction whose denominator is 77, multiply the denominator (11) by 7 as 11 x 7 = 77.
To not change the fraction, if we multiply the denominator by 7, you also multiply the numerator by 7 which gives:

6 / 11 = (6 x 7) / (11 x 7) = 42 / 77

d/ To express 7 / 10 in a fraction whose denominator is 100, multiply the denominator (10) by 10 as 10 x 10 = 100.

To not change the fraction, if we multiply the denominator by 10, you also multiply the numerator by 10 which gives :

7 / 10 = (7 x 10) / (10 x 10) = 70 / 100.

Exercise 14

To compare fractions, keep in mind the rule that we saw:

To compare two fractions, always put the same denominator.

When two fractions have the same denominator, the largest of the two is the one with the largest numerator.

a/ To compare 7 / 8 and 2 / 3, put the same denominator. We must find a common denominator to 8 and 3. 24 is a common denominator as 8 x 3 = 24 and 3 x 8 = 24. Remember that when you multiply the denominator of a fraction by a number, you have to multiply the numerator of the fraction by the same number; otherwise you will change the fraction! It gives:

7 / 8 = (7 x 3) / (8 x 3) = 21 / 24
2 / 3 = (2 x 8) / (3 x 8) = 16 / 24

Now the two fractions have the same denominator so the largest of the two is the one with the largest numerator, that is why 21 / 24 is larger than 16 / 24. We deduce that 7 / 8 is larger than 2 / 3.

b/ Similarly, the common denominator to 12 / 5 and 41 / 15 is 15 as 5 x 3 = 15 and 15 x 1 = 15, so:

12 / 5 = (12 x 3) / (5 x 3) = 36 / 15
41 / 15 = (41 x 1) / (15 x 1) = 41 / 15

As 41 / 15 is larger than 36 / 15, then 41 / 15 is larger than 12 / 5.

c/ The common denominator to 6 / 7 and 18 / 21 is 21 as 7 x 3 = 21 and 21 x 1 = 21, so:

6 / 7 = (6 x 3) / (7 x 3) = 18 / 21
18 / 21 = (18 x 1) / (21 x 1) = 18 / 21

We deduce that the two fractions are equal: 6 / 7 = 18 / 21.

d/ The common denominator to 5 / 9 and 7 / 11 is 99 as 9 x 11 = 99 and 11 x 9 = 99, so:

5 / 9 = (5 x 11) / (9 x 11) = 55 / 99
7 / 11 = (7 x 9) / (11 x 9) = 63 / 99

As 63 / 99 is larger than 55 / 99, then 7 / 11 is larger than 5 / 9.

Exercise 15

Remember the rule to add fractions:

To add two fractions, they must first be put to the same denominator.

130

Then we add the numerators together. The denominator remains unchanged.

So (A / B) + (C / B) = (A + C) / B

a/ To add 4/5 + 1/3 + 7/30, start by finding a common denominator to the three fractions : which number is common to 5 ; 3 and 30 ? To answer, find a number which is in the multiplication table of 3, 5 and 30. The answer is ... 30, as 5 x 6 = 30 (table of 5), 3 x 10 = 30 (table of 3) and 30 x 1 = 30 (table of 30).

4/5 = (4x6) / (5x6) = 24/30
1/3 = (1x10) / (3x10) = 10/30
7/30 = (7x1) / (30x1) = 7/30

So 4/5 + 1/3 + 7/30 = 24/30 + 10/30 + 7/30 = (24 + 10 + 7)/30 = 41/30

b/ Similarly to add 2/11 + 3/2 + 13/3, start by finding a common denominator to the three fractions : which number is common to 11 ; 2 and 3 ? To answer, find a number which is in the multiplication table of 2, 3 and 11. The answer is ... 66, as 2 x 33 = 66 (table of 2), 3 x 22 = 66 (table of 3) et 11 x 6 = 66 (table of 11).

2/11 = (2x6) / (11x6) = 12/66
3/2 = (3x33) / (2x33) = 99/66
13/3 = (13x22) / (3x22) = 286/66

So 2/11 + 3/2 + 13/3 = 12/66 + 99/66 + 286/66 = (12 + 99 + 286)/66 = 397/66

Exercise 16

Remember the rule to multiply fractions:

*To **multiply two fractions**, multiply the numerators together and multiply the denominators together.*

So (A / B) x (C / D) = (A x C) / (B x D).

a/ 3/5 x 6/5 = (3x6) / (5x5) = 18/25
b/ 7/11 x 9/4 = (7x9) / (11x4) = 63/44
c/ 3/14 x 5/2 = (3x5) / (14x2) = 15/28

Exercise 17

Remember the rule to simplify fractions:

A fraction can be simplified if it is possible to decompose the numerator and denominator to display a common number.

For example, the fraction A / B can be simplified if it is possible to write it as (axc) / (bxc).

The simplified fraction will be a / b.

And there will be A / B = a / b.

a/ To simplify 12/36, try to write 12 as a multiplication and 36 as a multiplication, in these two multiplications you must have a common number.
For example 12 = 12 x 1 and 36 = 12 x 3 (there is 12 in the two multiplications) so:

12 / 36 = (12x1) / (12x3) = 1 / 3

b/ Similarly to simplify 56/88, try to find a number which is in the multiplication tables of 56 and 88, it is the case for 8 as:
56 = 8 x 7 and 88 = 8 x 11

So 56 / 88 = (8x7) / (8x11) = 7 / 11

c/ For the fraction 9/14, you cannot find a number which is in the multiplication tables of 9 and 14. That is why this fraction cannot be simplified.

Exercise 18

I have a pie:

a / If I eat 5/9, we consider that the pie was cut into nine shares. Initially, nothing was eaten so the pie is full with nine shares present on 9 cut parts. The whole pie is 9/9. At the end, it remains 9/9 - 5/9 = (9-5) / 9 = 4/9.

b / If I eat 11/14, we consider that the pie was cut into 14 shares. Initially, the whole pie is 14/14. At the end, it remains 14/14 - 11/14 = (14-11) / 14 = 3/14.

c / If I eat 7/16, we consider that the pie was cut into 16 shares. Initially, the whole pie is 16/16. At the end, it remains 16/16 - 7/16 = (16-7) / 16 = 9/16.

Exercise 19

Remember the rule we discussed in the book:

To take a fraction of a quantity, multiply the fraction and the quantity.

a/ 3/4 of 5/8 means 3/4 x 5/8 = (3x5) / (4x8) = 15/32

b/ 1/6 of 7/10 means 1/6 x 7/10 = (1x7) / (6x10) = 7/60

c/ 4/5 of 3/4 means 4/5 x 3/4 = (4x3) / (5x4) = 12/20

This last fraction can be simplified as 12 = 4 x 3 and 20 = 4 x 5
So 12/20 = (4x3) / (4x5) = 3/5.

Exercise 20

Remember that the square of a number is that number multiplied by itself, as follows:

a/ 6^2 = 6 x 6 = 36
b/ 8^2 = 8 x 8 = 64
c/ 11^2 = 11 x 11 = 121
d/ 14^2 = 14 x 14 = 196

Exercise 22

Remember:

The square root of A is when the number multiplied by itself gives A

a/ square root of 25 is 5 as 5 x 5 = 25
b/ square root of 1764 is 42 as 42 x 42 = 1764
c/ square root of 441 is 21 as 21 x 21 = 441
d/ square root of 289 is 17 as 17 x 17 = 289

Exercise 22

To write a number as a power of 10, you must first determine if starting from the number 1, 0 is added to the right, in this case the power is positive, or if we add 0 to the left then the power is negative:

a/ 100 : starting from 1, we write 2 zeros to the right, so $100 = 10^2$ (positive power with 2 zeros).

b/ 10 : starting from 1, we write 1 zero to the right, so $10 = 10^1$ (positive power with 1 zero).

c/ 0 cannot be written with a power of 10, because it does not contain digit 1.

d/ 1 : starting from 1, we write 0 zero to the right, so $1 = 10^0$ (positive power with 0 zero). Any number to the power of 0 equals 1 ($2^0 = 3^0 = 4^0 = 25^0 = 45^0 = 1$).

e/ 10000 : starting from 1, we write 4 zeros to the right, so $10000 = 10^4$ (positive power with 4 zeros).

f/ 10000000: starting from 1, we write 7 zeros to the right, so $10000000 = 10^7$ (positive power with 7 zeros).

g/ 0.1 : starting from 1, we write 1 zero to the left, so $0.1 = 10^{-1}$ (negative power with 1 zero).

h/ 0.0001 : starting from 1, we write 4 zeros to the left, so $0.0001 = 10^{-4}$ (negative power with 4 zeros).

i/ 0.00000001 : starting from 1, we write 8 zeros to the left, so $0.00000001 = 10^{-8}$ (negative power with 8 zeros).

Exercise 23

To multiply with powers of 10, we apply the following rule:

A^n x A^m = A^{n+m}

a/ 10^3 x 10^4 = $10^{(3+4)}$ = 10^7
b/ 10^{-9} x 10^{-5} = $10^{(-9-5)}$ = 10^{-14}
c/ 10^{-4} x 10^6 x 10^{-1} = $10^{(-4+6-1)}$ = 10^1
d/ 10^{-6} x 10^{11} x 10^2 = $10^{(-6+11+2)}$=10^7

Exercise 24

We learnt:

To divide a number by A^n you just have to multiply this number by A^{-n}.

To divide with powers of 10, we deduce the following rule :

A^n / A^m = A^n x A^{-m} = A^{n-m}

a/ 10^5 / 10^2 = 10^5 x 10^{-2} = $10^{(5-2)}$ = 10^3
b/ 10^{-2} / 10^{-5} = 10^{-2} x 10^5 = $10^{(-2+5)}$ = 10^3
c/ 10^{-4} / 10^6 = 10^{-4} x 10^{-6} = $10^{(-4-6)}$ = 10^{-10}
d/ 10^{11} x 10^{-6} = 10^{11} x 10^6 = $10^{(11+6)}$ = 10^{17}

Exercise 25

Remember:

Scientific notation is the writing of a number as a decimal number with one digit from 1 to 9 before the decimal point, and multiplied by a power of 10.

a/ To write 0.54875 as a scientific notation, you have to write 5.4875 (decimal number with one digit from 1 to 9 before the decimal point) multiplied by a power of 10.

To go from 5.4875 to 0.54875 just shift the decimal point 1 digit to the left, that means 10^{-1}.

So $0.54875 = 5.4875 \times 10^{-1}$.

b/ To write 0.00054699 with a scientific notation, you have to write 5.4699 (decimal number with one digit from 1 to 9 before the decimal point) multiplied by a power of 10.

To go from 5.4699 to 0.00054699 just shift the decimal point 4 digits to the left, that means 10^{-4}.

So $0.00054699 = 5.4699 \times 10^{-4}$.

c/ To write 15488745 with a scientific notation, you have to write 1.5488745 (decimal number with one digit from 1 to 9 before the decimal point) multiplied by a power of 10.

To go from 1.5488745 to 15488745 just shift the decimal point 7 digits to the right, that means 10^{7}.

So $15488745 = 1.5488745 \times 10^{7}$

d/ To write 14589.547 with a scientific notation, you have to write 1.4589547 (decimal number with one digit from 1 to 9 before the decimal point) multiplied by a power of 10.

To go from 1.4589547 to 14589.547 just shift the decimal point 4 digits to the right, that means 10^4.

So $14589.547 = 1.4589547 \times 10^4$.

Conclusion

100 pages, this is the challenge that we set at the beginning of this book to make you understand decimal numbers, relative numbers, squares and square roots, fractions and powers through examples of everyday life.

I hope that by the time you close the book, you will know more about mathematics than when you opened it and especially that you took pleasure in reading these pages. If some concepts do not seem clear, take the time to read the related chapter.

If this learning convinced you, thank you for taking time to leave a positive review of this book on the website where it is offered for sale, and recommending that book to those around you. If, unfortunately, this book was not up to your expectations, do not hesitate to send me your comments by email, (pascal.imbert@yahoo.com) suggesting clear improvements that you think will be interesting to make. These elements are very important and will greatly help to enrich this book over the revisions.

The connection and exchange between reader and writer through feedback is essential for everyone's progress: enabling readers to heal their allergy to math, and the author towards the satisfaction of providing an effective remedy for this allergy...

www.ingramcontent.com/pod-product-compliance
Lightning Source LLC
Chambersburg PA
CBHW072254200526
45168CB00016B/1889